ON THE WINGS OF EAGLES
A Flight Instructor's Tale

James Atkinson

&

Fred Atkinson Jr.

On The Wings Of Eagles © 2013 by James B. Atkinson

All Rights reserved. No part of this book may be used or reproduced in any manner whatsoever, including Internet usage, without written permission from James B. Atkinson except in the case of brief quotations embodied in critical articles and reviews.

Cover Photographer: James B Atkinson and Rushpoint Stock
Cover Design: Courtney Atkinson
Editing and Layout: Courtney Atkinson.

ISBN: 9781729432754

DEDICATION

Thank you to my daughter, Courtney Atkinson, for all her hard work in editing this manuscript and her computer skills. Without her hard work this project would never have been finished.

CONTENTS

	Prologue	i
1	The Beginning	1
2	Barnstorming	13
3	Flyboys in WWII	17
4	Adventures with John	29
5	Hondo and the Korean War	51
6	The Farm and Fort Leonardwood	79
7	Hondo II	87
8	Missouri Bound	96
9	Vichy Airport	122
10	St. Clair Airport	134

PROLOGUE

An era in this world is fast coming to a close - the era in which man soared into the heavens in search of himself and set himself apart from the world below.

Today flying is commonplace. The average person can take a flight on an airliner for business or pleasure without a second thought of the sacrifices made in aviation. The aviator in the open cockpit biplane straining to see through the grease stained goggles looking for a field below has disappeared. His knowledge that man was not meant to fly was overcome by the fact that he is alone soaring with the eagles.

The facts in this book describe a man who has dedicated his life to flying. A man like most pilots who loves their craft and continues to pass it on like all wise teachers since the beginning of time. Those men are heroes too.

They may not break speed records, distance records, climb to the stars in the space shuttle, or fly fighters, but they are the beginning. They are instructors. Without instructors, and the sacrifices they made, the advances and conquests of flight would not be possible.

This account is about one man who lived his whole life doing what he enjoyed; flying and instructing. His philosophy for life was to learn from yesterday but not to live in it. Live for today, and plan for tomorrow.

This man actively flew for 56 years. His love of flying began as a barnstormer in the 30's and continued through the 80's. He was not a war hero or a celebrity, but he and so many more like him, were the glue that held aviation together. From a barnstorming young aviator to a flight instructor, he influenced many who later became participants in one of the greatest activities in the world.

The man described in this book is real. The stories are true. I know because he is my father. However, this book is dedicated to all who fly.

Men Who Fly
By Pat Branin Douglass

The men who fly are a breed of men
Unto themselves. We'll not know again
The little boy on a grassy hill
Who sees a hawk and knows the thrill
Of the summer wind on an upturned wing,
And the joy a graceful flight can bring.

There was a dream in this boy's eyes
That reflected the challenge of distant skies.

The passing of time and the graying of hair---
But the eye is still sharp and the light is there.
And he sees as he scans the far blue sky,
A dream that is missed by the passerby.

The men who fly are a special breed.
It's true—they spring from a certain seed.

A new kind of pilot has now made the scene;
His flesh is firm and his mind is keen.
He's good--it's true—no need to ask.
The computers say he's right for the task.
His eyes, like steel, his determined face
Show he's looking farther into space.

But his life will never know the thrill
Of the little boy on the grassy hill.
Where, as far as his eager eyes could see,
The air was clean and the sky was free;
Where the hawk soared high on the summer air
And the boy imagined he was there.

Before it's too late, if the world is wise,
It will honor these men who love the skies.

1 THE BEGINNING

I killed my father. I don't mean that I put a gun to his head and pulled the trigger. My stubbornness was more than he could handle. You see, I was not the wildest kid in Longview, Texas, growing up, but I could sure get in trouble. It was the Great Depression of 1929, and I was one of the few kids with a little money.

In 1928, my father was invited to Longview to open a pickle factory. He had already done that in Ada, Oklahoma, and he felt that Longview would be a better place. So we went.
Before opening a factory, Pappa had been a teacher in Ames, Iowa. He had come to Iowa from Ohio. He had rode a bicycle all the way, until he settled in Ames. He was so tough I would have hated to be one of his students. He wouldn't even let me have a bike. He thought the kids were a nuisance to automobiles.

I had one of the hottest cars in Longview. I earned my money working at the pickle factory, and I worked hard for my money. I never stole anything. I didn't have to steal. I would just speed around Longview.

Longview had brick streets downtown where my friends and I would drive around until we saw a brick dislodging itself from the street. I'd drive over it just right so the bricks would fly up. If we were lucky we'd break a window out of one of the stores.

Eventually, the time came for me to go to college. Pappa picked out Iowa State, in Ames, for me. It wasn't that Texas didn't have good schools; he wanted me out of town. After I got to Ames, I did my studies diligently. That is, until a barnstormer came to town, and I got a ride.

It was then that I found what I wanted to do. I was hooked.

I would go riding all the time. I didn't think I could learn to fly; they didn't have many flight schools around in 1929. But I kept paying the barnstormers, and they kept taking me up in the air. Every time a pilot would come to town I would go to the local field just to catch another ride.

After the first year of college, I went back to the factory and worked. You see, it was my job to clean out the pickle vats to get them ready for a new batch. Pappa didn't want to hire someone to do that because of all the cleaning chemicals it took to clean them. So I got the job. That summer was not the easiest of my life, 12 hours a day for 90 days.

I decided to go back to school. This time, I went to Oklahoma University. At least it was closer to home and closer to a lot of flying. I spent all my spare time at the airport. They had a real one. I would wash airplanes, work on them; generally do anything just to get a ride.

I flew with a lot of different people in those days, and I started to get the idea I might be able to fly one of those things.

By the end of the year, it was time to go back to Longview.

Pappa saw how the chemicals were affecting me so he decided to come help me. I was, and am, stubborn, and Pappa wanted to prove that he could still outwork his son. So, we started cleaning those vats, and cleaning, and cleaning, until well after dark. Neither one of us willing to give in.

Pappa collapsed about 10:00 p.m. He went to the hospital and never came out.

I wasn't feeling well either and went to the doctor. The doctor said I had burned my lungs with the chemicals and highly suggested I change my career, or I would end up like Pappa.

I knew my pride and stubbornness of wanting to better him, killed him. After the funeral, even though I was the only son, I didn't want anything to do with the pickle factory. So, one of my sister's husbands ran it.

My mother was left in very good financial shape when Pappa passed away. This was the early thirties when a dollar was a dollar. So I had financial freedom.

I figured I might as well fly for a living since I

probably wouldn't live very long anyway.

Little did I realize, my decision would set the stage for the rest of a very long life.

After flying in Ames and Oklahoma City, I was convinced that I wanted a new airplane fresh off the factory floor. There were all kinds of airplanes for sale at that time, but I picked a Franklin Sport biplane. It had a ninety horsepower Lambert radial engine and was small enough I could put in a barn.

At 20 years old, I rode the train to Franklin, Pennsylvania, to tour their factory and buy an airplane. It turned out Franklin only built 12 airplanes before going belly up during the depression. I had accidentally secured myself a limited edition airplane.

My trip to Pennsylvania was uneventful. The nice scenery and club car poker kept me occupied.

After arriving at Franklin, I made my way to the factory, introduced myself, and proceeded to negotiate for an airplane. After deciding on a price, (all of $600) the salesman told me it would take six months after the order was placed for it to be ready to be delivered.

Six months was a lifetime.

After much debate, he conceded that one was built, but it waiting for the buyer to come up with the money. The salesman promised I could have the plane (since I had the cash) if the prospective buyer didn't show up with the money in three days.

Those three days passed very slowly. On the morning of the fourth day I called the salesman. He said if I still wanted it, the plane was mine. I dropped the phone, exploded out of the hotel, grabbed a cab, and was at the factory before the salesman hung up – almost. We took care of the paperwork. I paid the price. And we made our way to the flight line where my new Sport was waiting.

The factory pilot briefed me on the airplane for about an hour, going over the eccentricities of the plane. He then asked me if I had any questions.

"Just one," I replied, "how do you fly it?"

I had been up several times but didn't know anything about flying or navigation. It was a long way back to Texas, and there weren't any road signs. After a shocked look and five minutes of laughter, we struck a deal. I would pay him $50 and a train ticket back if he would instruct me on the way back to Longview. It is still to this day, the best deal I've made in my entire life.

We took off early on a beautiful Friday morning. The air was heavy, and there was a slight breeze. The day before we had gone over the fundamentals of flying, so Bill (the instructor/test pilot) let me takeoff. I taxied until I was on the runway, a grass field, and placed the plane into the wind. I revved the engine. The plane started forward. There were no brakes. I applied the right rudder for torque correction, and we were off!

My first flight in my first airplane! It would be another 33 years before I owned another one.

Between the weather, forced landings, trying to find gas, and barnstorming a little, we finally made it to Hot Springs, Arkansas, on April 1st, 1931. Bill decided I had had enough instruction, and he soloed me at 5:00 PM. He hopped on a train back to Pennsylvania. I spent the night and flew on to Longview the next morning. Once in town, I made arrangements to use a field on the east side of town that eventually became The East Side Airport.

Theoretically, I had the first airplane based in Longview.

By the time I reached Longview, word had spread. It seemed like the whole town came out to see the airplane and take a ride. I probably took up ten people for their first ride in an airplane my first day back in town.

I didn't have to have a license, insurance, or even an airport! That's the way it was in 1931. Licenses were available, but most pilots just flew around until the government made them get licenses.

I flew around Longview and the surrounding areas for a few months. I accumulated a lot of experience; good, bad and hysterical.

There was no starter on the engine, and no brakes on the wheels. It always helped to have a little help with chocks when I had to prop (hand start) the engine. Usually someone was around but occasionally there wasn't. On one occasion, I set the engine up to start one day and forgot to set the chocks. I was all by myself and figured I could do it.

Unfortunately, I had the throttle cracked too

much. When I pulled the prop through, it started and revved up too much. Then the plane started rolling forward, and I wasn't fast enough to run around the plane to pull the throttle back. But, I did manage to grab the wing tip on the left side.

Here I am, holding on to the wingtip, and I'm spinning around trying to keep the airplane from running away from me. This went on for about 3 revolutions. After far too many minutes and a few panicked choice words, I managed to work my way to the cockpit to pull the throttle back.

Whew, did I feel stupid, and I never tried to start the airplane without the chocks on the wheels, or me being in the cockpit. This particular event caused me to think that there must be a way to start the engine from the cockpit and make it safer for me and the airplane. I always tinkered with stuff and engines in particular. That's why I had the hottest car in Longview.

Airplane engines get their fire for the sparkplugs from a magneto instead of a distributor like on a car. I came up with a device. It was a generator of sorts, and I hooked it up to the number one spark plug. I had the generator itself in the cockpit with a handle to turn. When I turned the generator handle, it would produce enough electricity to shower the cylinder with sparks and if the number one cylinder was on top - dead center - and there was fuel in the cylinder it might fire enough to turn the engine over and start. It was very crude, but it worked enough to make it worthwhile.

After a while, I got tired of people coming

out to watch me and ask for free rides. I found a mean, but effective, way of getting people to leave me alone. I had hooked up my generator to the fence. People would be hanging around and leaning on the fence. I would act like I was getting ready to fly by getting in the cockpit. Then I would spin the generator. The electricity would flow to the fence, and all the people would jump back after getting the message.

Few of my adventures ever proved dangerous; however, one I remember in particular almost got me killed. I was up practicing aerobatics hoping to get into an air show group.

While practicing loops and spins I happened to notice a truck parked on a remote oil field road. Two guys were busily unloading crates and hiding them behind pine trees beside the road. After they left, I figured what's left alone belongs to anyone. So, I took a look at the clay road and decided it was suitable for a landing.

After some coaxing I landed and taxied up to the pine trees. Not knowing what to expect, I parked on an incline, throttled back to idle, and got out. I always took a set of chocks for situations like this and placed them in front of the wheels. Feeling the airplane was secure, I went off to inspect the crates, and much to my surprise there were forty cases of bootleg whiskey. I hit the mother-load! Since it was difficult to put forty cases in the cockpit, I opened one of the cases and stored bottles everywhere I could in the cockpit.

After placing a few bottles in the cockpit, I

heard car engines close by. I decided there was no sense in tempting fate, so I pulled the chocks and started taxiing for takeoff. During my haste to get airborne, I hit a small chuck hole which ordinarily would be no problem except one of the bottles rolled up under the left rudder peddle and lodged itself in place. While I stopped to remove the bottle, the car drove up to the hidden booze to investigate why an airplane was on site. The men carried the biggest guns I'd ever seen.

I hurriedly dislodged the bottle, jumped back in the cockpit, and prayed. I was never a religious man so I figured today would be as good of a day as any to form a relationship with God. As I began my takeoff roll I could hear shouting and shooting. I know they weren't squirrel hunting. I just barely cleared the trees and the curve in the road. I kept flying all the way to Dallas and didn't look back.

When I landed at Love Field, I taxied to Colonel Long's Hangar and pulled inside. I was too far from the bootleggers for them to have gotten my number, but not too far to receive six neat little holes behind the cockpit.

Needless to say, I sweated blood for a few weeks, but the whiskey was good!

Pilots tend to be their own breed of people. Colonel Long, a pilot in WWI, or so he said, held Sunday services in his hangar at Love Field. I would fly in from Longview to spend the morning with a group of ever-changing pilots and mechanics. We all liked him enough that we didn't investigate the wild stories he told like folks do now-a-days

Our kind of Sunday service suited us. We would have a crap game on the floor of the hangar. Always gambling and telling stories.

One Sunday, I got in early before the storms moved in. We played for a while, but I ran out of money and decided to fly back to Longview. It was raining cats and dogs, and I had an open cockpit biplane, mind you. Not even the Colonel wanted to get wet pushing my plane out of the hangar. I asked if he would open the doors. That he could do, but just enough to get my plane out.

I started the engine, with wheel chocks in place, and the Colonel opened the doors just wide enough for me to rev up while someone else pulled the chocks. I was airborne just after I left his hangar. Once I cleared the hangar I looked back, and they already had the doors closed.

I spent the spring, summer, and fall barnstorming around Arkansas, Louisiana, Oklahoma, and Texas. Nearly always, I tried to fly downwind. I got there faster. By 1932, airplanes were more common, and it was difficult for a barnstormer to make any money. Luckily, I had my money from the pickle factory that kept me going. During this time, I would fly over a little town and circle and try to get people's attention. If people came out and looked up to see this crazy fool, I would find a field close to town and land.

The farmer for sure would come out. I would strike up a deal with him to use his field. More often than not, I'd be invited to supper and have a bed to sleep in.

One day, I was in rice country in Arkansas when I encountered engine problems.

Naturally, I decided to land in the closest available field. Unfortunately for me it was a rice field. I landed and probably didn't roll but 10 yards in the mud. As I was climbing out of the cockpit the owner of the field came up.

He asked the usual questions.

"Was I alright?"

"Any Problems?"

"Can I help?"

"Have you had lunch?"

Sucker! I mean the nice man invited me to lunch. How could I say no?

He was a pleasant sort. He helped me tinker with the engine, and within 30 minutes, I had the engine running. You see, I was such an expert mechanic, I could go to the engine to twist this, hit that, kick a little, curse a little, and the engine would start! Nowadays we know the problem is carburetor ice. But in those days we didn't know about it. We did know that if we tinkered with the engine for a few minutes and waited for the ice to melt, the engine would start.

I was ready for takeoff. I thanked the farmer for the meal and offered to pay for the rice I would ruin on takeoff.

The unknowing farmer replied, "You didn't tear anything up on landing. Why would you on takeoff?"

Little did he know that a 90 horsepower engine pulling a biplane in the mud takes some distance to accelerate to flying speed. It took three quarters of a mile and five rows of rice to get airborne. I never looked back. I was afraid the old farmer was going for his shotgun!

The Franklin Sport only had enough gas to take about 15 people on rides before needing more gas. Luckily, there was always a kid who was willing to take my gas can and go back into town for gas if I'd give him a free ride. More than once, it was usually the best ride of the day. And then, after spending the night with the farmer and his family, I'd be on my way to the next town.

Sometimes, people wouldn't come out and wave at me when I flew over. I figured somebody had already barnstormed their way through there. I would just fly on to the next town.

I did this for about a year. When I got home, I only had about $10 in my pocket and not much was left of the pickle factory. My money was running out, and my barnstorming days looked over. I stored the plane over the winter and got a call from Bailey Air Shows to come fly with them starting in April.

That just made my year! I could fly, have fun, and make a little money. I still didn't have a license, but nobody seemed to care, so I just went on flying.

Talk about a child's dream of running away to the circus. But this was my kind of circus. And, as it turns out it wasn't just my kind of circus, but my future son's circus too. But, that's getting ahead of ourselves.

2 BARNSTORMING

Bailey Air Shows brought an order to my flying that had been lacking which helped me get better at flying, but then so did simply flying with other air show pilots for a change. We had three aerobatic airplanes, a parachute jumper, and a Curtiss Robin for airplane rides. We would go from town to town just as I had before. But this time when attendance dwindled we would come up with something more exciting each time. Each town would receive sightseeing flights and "death defying" stunts that looked far more exciting from the ground then it did in the air. But it sure was fun.

When we came into town, we would wing-walk down the streets. Which meant we would fly with one of our wings pointed down to the road while the other wing pointed high into the sky. It felt like an acrobat in the air. We would drop flyers and even fly upside down through Main Street. We'd do just about anything to attract attention.

Bailey did make me get my private license eventually. I got it on June 1st, 1934. I met with the CAA man. He observed me from the ground as I completed flight maneuvers he told me to do when we were on the ground together. I guess I did alright; my number is 20272. Which means I was the 272nd pilot in the southwest. They gave each region of the country 20,000 numbers and figured there would never be more than a million pilots. Boy, were they wrong!

Although we were in a dangerous occupation, many humorous events happened. One which comes to mind concerns a crusty old parachutist named Sam. While we were performing in a small town in Texas, Sam broke his arm making a quick getaway from a bedroom window. He knew that the crowd at the air show the next day would want a parachutist, so he looked for a volunteer.

A young man caught up in the glory of flight stepped up and said, "Yes Sir!" However, he had never jumped much less been in an airplane before. Well, Sam did the best he could to instruct the poor fellow on the ground. But time ran out. So Sam, myself, and the brave volunteer got in the Robin and took off.

This airplane had a pilot's seat and small cabin for the other two. As I proceeded to gain altitude I could make out Sam's attempt to instruct the brave soul, but I could tell the higher I got the paler he got.

We made our pass over the field the kid was

supposed to jump over. I glanced back and almost had a heart attack. The kid was in the doorway facing the cabin with both hands on the edge of the doorway screaming "no" at the top of his lungs. Needless to say, this is not how you are supposed to parachute out of a perfectly good airplane.

Sam was sitting on the floor facing the kid and trying to force him out of the door. Sam kicked him. Sam hit him. Sam screamed at him. And, finally the kid fell out of the door screaming for God's help.

I knew we would be charged with murder. I even debated with myself on heading south to Mexico instead of landing. But, my wits got the better of me. I reduced power and started to follow the kid down. Sam's instruction must have been good, because he pulled the rip cord and awkwardly landed. I was afraid he was going to skin us alive.

I ran out of the Robin and into the crowd in hopes of avoiding the soul we violently pushed out of the plane. Finally, Sam and I got the nerve up to go back to the air show tent, and there was the kid. He ran up to us and hugged us and asked Sam if he could do it again.

Before Sam could answer I said "No!"

My part of the air show was to do 15 minutes of exciting aerobatics. I learned real quick that I had to keep coming up with new maneuvers. Well, the Sport only had a 90 horsepower engine. There was a limit to what it would do.

It did spins and loops well, but inverted maneuvers were really limited. My signature maneuver was a loop on takeoff. I had to accelerate up to 100 mph to start a loop so I would keep the plane very low on acceleration. When I hit 100 mph, I would start my loop. Of course, I had to start a quarter of a mile from the loop starting point. This went on for most of the summer but as the summer wore on, the temperature rose and my looping ability became less and less.

Finally, in late July, 1934 I tried to loop in front of the crowd and came out of the loop, one foot too low, knocking out the landing gear and wrecking my beloved Sport. The only thing I hurt was my pride, but the Sport was badly broken.

Nowadays, we know that as the temperature rises, so does the density altitude which means the engine has less power, and the wings develop less lift for the increase in temperature. I got it repaired, and sold it to a lawyer in Ft. Worth.

In the mid-1980s, my son Jim saw an advertisement in Trade-A-Plane for a Franklin Sport in Ft. Worth. My other son, Fred, Jr. lived in Arlington, TX, so we decided to visit this particular airplane.

That old plane sure did look familiar. We looked it over real good and even found my old "hand generator." So it must have been mine.

It turns out this "Sport" had been in storage for more than forty years. The lawyer wanted to rebuild it. He contracted with a mechanic to restore it. There is some kind of lawsuit going on for years after it was restored. We tried to buy it, but with the lawsuit and crusty old guys who didn't want to give it up, it would've been years. Knowing that it was originally my plane wouldn't have mattered a bit.

3 FLYBOYS IN WWII

After I had the airplane repaired, I did not have the money or the inkling to return to the air show business. I decided to sell the Franklin Sport and concentrate on making money for a while. The only thing I really knew how to do was make and sell pickles. So, I worked for Heinz for a while, and I moved around a lot, from Sherman to Big Sandy and all points in between.

This went on for a couple of years, but the bug for flying was always in the back of my mind.

I was around Love Field in Dallas where I met my future wife. She encouraged me to fly again. I guess she liked the people and the notoriety that flying brought both of us. So, I decided to get my commercial license and made an appointment with a CAA inspector. The inspector I signed up with was a little shady come to find out. I would make an appointment, and he wouldn't show up.

This went on for about six months. I refused

to grease his palm and he refused to ride with me until I did. I finally relented and paid the exorbitant sum of $200, in 1938. He didn't even ride with me. He just gave me the license. So, I was off and flying for money now.

I started instructing at Love Field where I realized I really wanted to go back to Longview. By this time, East Side Airport had been in business for a while, and they needed flight instructors and charter pilots.

The oil boom of East Texas was in full swing, and everybody had money to go places and to learn to fly. I got a lot of experience in Curtiss Robins, Fleets, Stearmans, Wacos, and Great Lakes.

During this time, if I had an airplane somebody else wanted to fly, he would offer his as a trade, nobody worried about checkouts, or if the other pilot was competent. If he flew in, then he could fly out!

Then something earth shattering happened to my wife and myself. My wife, Nan, was pregnant. Well, I was 28, and, I guess, I was old enough to have a child. Now, I really had to find work if I wanted to fly.

In the middle of 1940, the US Army Air Corps was advertising for flight instructors to teach cadets in their airplanes, and they paid a monthly salary which was just what I needed.

The catch was that I had to travel to Randolf Field in San Antonio to go to school for three months. I met many other pilots from all over the country who were in similar situations to me and others who just couldn't learn how to teach.

The funny thing about the military training during the years of WWII is that the flight instructors of that era really learned how to fly. Before, we (the instructors who didn't know that much about flying) were teaching people who were lucky enough to learn to fly simply by getting into the air.

During the war the flying community really started learning the intricacies of flight. Precision maneuvers were introduced to bring all pilots into standardization. Randolph was a great learning experience for all of us, not only did we learn what the military wanted us to teach, but we also learned discipline in our lives and our flying.

We thought we pretty much knew how to fly, but in that three months we developed into real pilots instead of just barnstormers playing around. We didn't know where we would end up instructing for the military until the last week.

I could have ended up in California, or Georgia, but unlike most of the military thinking, I got posted at Bonham, Texas, about 100 miles from Longview. When I got back to Longview, my wife had had our first son Fred Harold Atkinson, JR. As soon as I got back we moved to Bonham, just one of the many moves we had ahead of us.

Bonham had an Air Corps Base just out of town. We got paid the exorbitant sum of $50 a month and the luxury of 100 hours of flying a month. The airplane we flew was a Fairchild PT-19. The P stood for Primary and the T stood for Trainer. The military had many other different types of PT airplanes at the time. I guess they tried to divide up the airplanes they were buying between a bunch of manufacturers. They had PT-17(Stearmans), PT-22's, PT-26's, and many other types.

The PT-19 we had was a single wing, open cockpit airplane with an inverted 6 cylinder engine of 110 horsepower. The spar in the wing was wood and caused some serious problems later on.

The Ryan Company had to manufacture so many of these airplanes at such a quick rate that toward the end of the contract, they didn't have enough varnish to keep the spars weather proof.

Toward the end of the time I was at Bonham, we would get the new PT-19's and after a summer and a winter in Bonham's 110 degree heat and 30 degree cold, the spar would collapse. It wouldn't take a hard landing, usually just a grease job and boom, airplane in the middle of the runway.

We had many amusing incidents happen at Bonham.

We were fledgling instructors teaching fledgling pilots. Anything could happen.

One incident involves a highly temperamental, but excellent instructor named Shorty Jacobs. Shorty was known for his unconventional methods in motivating his students. We all had to come up with our own individual motivational methods, but his methods were particularly interesting. And, I can't say that they would be effective if I tried them myself.

It was a particularly warm morning in Texas with just a slight crosswind and we were practicing landings and takeoffs, touch and goes, with our students. We had four airplanes in the pattern.

We practice many takeoffs and landings to hone the students' knowledge, heighten their ability in the traffic patterns, and, of course, to learn how to land and takeoff.

With four airplanes in the pattern, it also teaches students teamwork, flying in loose formation, and assists in judging climbing and gliding distances. It is a very important part of learning to fly especially since many of these boys would end up in Europe flying in formations.

The students also had to correct for torque, trim, maintain different speeds, and maintain different attitudes. They didn't just learn how to land and takeoff. Each man, typically, has nearly all of the motions of the aircraft in a matter of five minutes.

I began watching a particular PT-19 since it wasn't following the proper procedure. I noticed this was the airplane that Shorty took off in. We didn't have radios at the time, and the flight instructor could only communicate with the student through the gosport, a particularly evil form of communication we'll get into later.

The PT-19 with Shorty in it was having difficulties. It was bouncing as high as a hangar and flying erratically. This made Shorty slightly upset, and he left the traffic pattern. Luckily for my student and I, it gave a perfect example of the incorrect way to fly.

I didn't notice him for about ten minutes, but when I did, he was about 3 miles away, and the airplane looked odd. I can't say exactly what made it look odd at that distance, but there was something off about the angle at which it was flying. I decided to fly over and see what was happening.

As we were flying toward Shorty's airplane we noticed that it was flying all over the sky. It didn't have a specific pattern to its drifts and turns. But as we got closer we could make out a figure on the wing. Could it be a Gremlin?

There was Shorty standing on the wing. He had unscrewed the control stick from the instructor's seat, and was using it as a pointer and punishment stick from outside the aircraft. Shorty was giving the student a little "hands on" flight instruction by beating him.

I was amazed the student didn't just roll upside down and cause Shorty to fall off. We had parachutes on, but he was too low to get it open. I could just read the incident report, "My instructor climbed out on the wing and fell off!"

Later after work, I met Shorty and some of the other instructors at our favorite watering hole. Funny how much drinking alcohol and flying go together.

My curiosity was aroused, so I confronted Shorty about his temperamental tactics. According to Shorty, this student had 10 hours of dual instruction and was no closer to soloing than the man in the moon (at that point it was typical for a student to receive 8 hours of dual construction before going onto the next phase of his training.)

Apparently, he always froze up on the final approach. His feet were uncoordinated. His handling of the stick was erratic, and, in general, his brain was mushy. It was a classic case of lack of confidence and a little fear of hitting the ground. Up until this student, Shorty's record was great and he had never pink slipped a student. Pink slipping a student means he flunks out and could go back for remedial training or be kicked out.

He told me that "no son-of-a-bitching student" was going to spoil his record. So, he was either going to beat him to death or teach him to fly.

It takes fear to conquer fear because that student shaped up and graduated. He admitted to Shorty later that he was afraid of flying, but after an incident with a mad man on the wing, he was more afraid Shorty would kill them both next time he screwed up. So, he made up his mind to conquer his fear of flying. He certainly couldn't change Shorty.

Bonham was in a dry county, they couldn't serve alcohol anywhere, and they certainly couldn't deliver alcohol to the officer's club. Once a month, we would have to drive to Dallas to pick up the supplies for the "O" Club. It was not really an officer's club, because we weren't officers. But it did allow us to drink in a dry county.

Bonham was about an hour and a half away from Dallas. So, we could generally test all of the supplies on the way home. More often than not, it would take four to five hours to get home. We were in an army truck. Which meant that the police would never stop us.

One day, in 1943, I was out soloing a student at an auxiliary field, a practice field set up for the aforementioned takeoffs and landings. I set my student up for his third supervised solo. He was to fly and do takeoffs and landings for 30 minutes. I got out of the back seat, told him what to do, and went over to a tree offering free shade.

My student took off and another instructor and his student come over to rest and talk under the shade of the tree. It was a popular spot. We talk about the war, girls, etc. My student eventually finished his 30 minutes, lands, and taxis over to us.

He shuts down and walks over. He doesn't say much and keeps staring at the other instructor. They leave and we start to talk over his flight when he interrupts me.

"Do you know who that was?"

"No, we were just talking," I said.

"That was Robert Taylor, the movie star."

Well, you could have fooled me. Of course, I wasn't too impressed with actors since I didn't go very often. You could have put Marilyn Monroe in front of me, and I wouldn't know her from the check out girl from the corner store.

A little while ago, I mentioned the gosport and said I would continue. Now is the time. The gosport was a set of tubes that ran from the students ears to the instructor's mouth. It is a lot like a stethoscope only longer. The instructor could talk, typically yell, into the instrument, and the student was supposed to hear the message and comply. There was no way the student could ask a question until they were on the ground. The gosports were one way communication. Often, the instructions were not clear and the student would do the wrong thing.

Some sadistic instructors found ways to "motivate" the students to hear better. One way was to hold the speaking part of the gosport into the wind, this would tend to blow the wind into the student's ears. And, ultimately, the student would know he did something bad. Eventually, some of the instructors who chewed tobacco modified the wind communication. They would spit their wad into the gosport and then stick it into the wind. This really made a mess. I hated getting into a plane after one of these motivational rides.

I had the only student who was a natural born pilot while I was here. Everything I taught him to do was always done once and that was that. He aced all the tests even though he had never flown before. I kept track and communicated with a lot of the students that graduated with me, and he was one of them.

He went on to fly B-24's in the European Theater, and on one of his first bombing runs over Ploesti, in Romania, during Operation Tidal Wave he was shot down and killed. That goes to show you that even the best pilots and the most studied pilots can get shot down. Those were your chances in WWII.

The PT-19 was about the most docile aircraft of the era. The BT-13 on the other hand was one of those aircraft that could turn around and bite you like a pit bull. The Basic students from

Sherman liked to find some of us in PT-19s and try to dogfight. When they got behind us, we would dive down to about 1000 feet and start turning. Since the BT was bigger and heavier, we could turn tighter than they could. We would roll into an ever tightening turn, and slow down. As the BT approached his stall speed he would increase his angle of bank. The pilots would try to cheat on the turn by pushing bottom rudder. Right after the pilot pushed bottom rudder, the BT would do the neatest little bottom rudder stall. After the BT recovered, we would fly up to the BT, because the student pilot always looked a little pea-ked. It was a sight to see.

Nowadays when a pilot does this, it is more than likely turning from base to final, where the pilot is trying to cheat on the turn, and the airplane doesn't have time to recover. This happens because the pilot is watching the runway instead of watching the flight of the airplane.

As I've told my students many times, "don't watch the runway. It's been there for years. It's not moving. We are."

Most of the civilian flight instructors at Bonham were within the age of being drafted, and there was no telling where the army would have sent us. So the army came up with the idea to voluntarily induct us into the army with the idea that it would keep us at Bonham as flight instructors.

Naturally, the army, being what it is, and working in its infinite wisdom, decided after we had been inducted that we were no longer needed at

Bonham. The army sent me to a multi-engine school in San Antonio to train for flying over the hump in Burma and India into China.

Through some fluke, or something, I was given a choice after I finished school. I could go ahead and fly the hump, or I could go to Wichita Falls, Texas, as a liaison instructor. Guess which one I chose.

I had a wife and small child by this time and did not really want to be separated from them. So, the first stop before Wichita Falls was Fort Sill, Oklahoma, for cub training. One of the instructors I had was an ex-liaison pilot who had flown in Europe. He was one of the few liaison pilots credited with downing a German Aircraft.

On our first orientation flight we were flying at about 500 feet looking over the countryside.

He said, "Would you like to see what we do when a German fighter gets on our tail?"

Being an expert in PT-19s, I figured what could this guy show me?

So I said, "sure."

We were flying over a field roughly 500 feet long. He rolled the Cub into a "Split S" (sort of a half roll and half loop aerobatic trick), landed, turned off the ignition, jumped out, and ran to the trees. By the time I got over the shock of doing a "Split S" at 500 feet, I was sitting in a Cub, with the engine off, and rolling to a stop. My instructor was standing at the edge of the trees, laughing.

From then on, I never let my guard down with this guy.

The commander of the field was an Ace in

Europe during World War II, and he also had a movie, Fighter Squadron, made about him in 1948. He and his wingman were flying in Europe in P-47s. His wingman got shot down. He landed, picked up his wingman, and they both shared the pilot's seat on the way back. So much for teamwork. When they got back to England, the press wanted to take pictures of how they did it. But they couldn't fit into the cockpit anymore! It was too small.

I learned more about slow flying, soft field takeoffs and landings, and, generally, flying a Cub than I could have in the civilian world. All this training would come back to me and help me later in life.

Not long after we moved to Wichita Falls the war was over. I was faced with a dilemma. How was I going to make a living with a wife and five year old child?

The U.S. was going to be overrun with military pilots coming back looking for flying jobs. I decided the best thing for me to do, in fact, it was the only thing I wanted to do, was to return to Longview, and instruct in the flight school at East Side Airport.

4 ADVENTURES WITH JOHN

It's funny the way one job leads to another. While I was instructing, I complained to one of my students that I wasn't making enough money as a flight instructor. But then, when did flight instructors make enough? John took lessons up until he soloed, at which time he decided he was more dangerous in an airplane as the pilot than having me fly him around. John, being the kind of person he was, always started out with conservative ideas.

He decided to hire me as a part-time pilot and full-time oil field inspector, meaning I was supposed to go around his fields and check on the gauges to make sure everything was all right. Funny thing was, I never did go out and inspect anything. We started flying to Galveston, Tyler, Dallas, Ft. Lauderdale, New York, California, Indianapolis, etc. Well, soon those big hubs just became a springboard for other adventures.

John decided he needed something faster

than the airplanes available to us so he decided to buy a brand new Beech Bonanza for the whopping price of $8,500. Nowadays, a fully equipped Bonanza sells for a half of a million! That's definitely a change.

All of a sudden, John had business and people he had to see all over the country. After we had the Bonanza for a while, John wanted to take the family to California. So we took off with John, Nell, John,Jr., and JD. As we flew West, John would point something out that he wanted me to fly over it.

It was summer time which meant that I wanted to fly high where it was cool, but John wanted the family to see everything. We spent the night in El Paso, Texas, which is about half way from Longview, Texas, to Los Angeles, California. We left early the next day and got to the Grand Canyon at about 11:00 a.m.

It was already about 110 degrees at two thousand feet above the ground.

"Let's get out the camera and fly down the canyon," John says.

Well, John got out the 8 millimeter movie camera, and I dove into the canyon at the beginning of the canyon. Talk about a rough ride! It was about 120 degrees on the outside air temperature gauge and bumpy as hell. Nell, John, Jr. and JD were throwing up in the back and John kept yelling that he wanted to go lower. When he would run out of film, I would pull up and plead with him to stop but he said keep going. So we flew the entire length of the Grand Canyon under the ridge at 150 miles per hour. John still has the film, and every time he got it out to watch it, the kids would throw up!

I was employed as a full-time pilot and flying all over the country. We even flew down to Mexico and over to the Bahamas. John liked to fish so he bought a boat. It was a forty foot express cruiser from Chris Craft. We caught many a fish off of that boat and broke a couple of fishing records.

The one fishing record I recall was for a sailfish we caught in Mexico on light tackle. When you fish for big fish on very light tackle, the boat operator is the one who actually is able to tire the fish out and all the fisherman has to do is reel in the fish. Two weeks after he secured the record, a 15 year old girl, with a broken leg, broke it.

The reason I remember this so well is that we decided to bring the carcass back to Longview to have it mounted. We wrapped it up, but good refrigeration in Mexico in the 1940's was non-existent. We flew into Nogales to cross the border. Nogales had the very best hamburgers I've ever eaten. So, we couldn't resist having lunch. Nogales in the summer is hot. Hotter than Texas hot. When we would call into U.S. Customs the restaurant would know and find out how many people were coming in.

Then when you landed, they would have the hamburgers waiting on the flight line before you cleared customs. The cokes were good. The hamburgers were American. Nothing better!

So, after clearing Customs and after trying

to convince the border patrol all we had was a carcass of a sail fish on board, they finally let us go. In the meantime, the carcass was rotting, and it smelled quite ripe. We still had another eight more long, hot hours to go. When we got back to Longview the smell was well beyond the realm of obscene. I never did get the smell out of that Bonanza. In fact, the linemen at the airport wouldn't put it in the hangar anymore.

We called Nell, John's wife, and asked her to pick us up. We smelled about as good as the airplane. So, she wouldn't let us in her Cadillac. I had to get the airplane and the carcass to Florida the next day to deliver the carcass to the taxidermist. I bought a case of air freshener and opened them in the airplane, spread it all over, and flew with the window open.

When I got to the taxidermist he said, "Why did you bring the whole fish? All I needed was the bill!"

I could've quit my job right then and there. But I was having so much fun. I flew back to Longview the next day. I went home, and my wife burned my clothes.

Even outdoors, the linemen still wouldn't et the Bonanza close to anything else. We had many experiences in that old Bonanza, it's a wonder we survived.

John liked to fish for tuna in the Bahamas but tuna are a migratory fish and sometimes hard to find. So John would go out in the boat and I would go flying. As I found the tuna, I would then fly back over John and point the way to the fish. We didn't want to get on the radio and broadcast the tuna's position, John did not want competition.

I do not approve of drinking alcohol around airplanes; you never know when a paying customer might show up.

Anyway, while I was with John, I took my one and only drink in an airplane. We were flying from Ft. Leonardwood, Missouri, to Pensacola, Florida, late one night as was John's custom. I had flown this route many times during the day and knew all the landmarks regardless of time of day. We were over a former Air Corps Base and the engine on the Bonanza quit cold.

I knew we were over this particular airport, but the only light on the entire airport at that time of night was an outdoor light by one of the hangars. We had been at 4,000 feet altitude. I set up a power off approach into the airport. A night landing is not easy power off, but without runway lights or landing lights on the airplane it was horrifying. All I could remember about the airport was the position of the light in relation to the runway but had never landed here before.

As we approached the runway, all I could use as a reference point was the all night light, 200 yards away. We made the approach, flare, and smooth landing. We rolled to a stop.

John said, "I need a drink!"
I said, "I'll take one too!"

We got out of the airplane after the drink, and the nose wheel was on the centerline of the runway. I had figured correctly. That was the only time since 1948 when an engine quit cold on me. And it was after 25,000 flight hours.

John was a kind hearted man and many times when we would have an extra seat, we would take along a friend or someone who wanted to go our way. Lawsuits were not as prevalent as they are today, so you wouldn't worry about taking someone. His generosity got us in trouble a few times but the worst one stands out particularly.

We were in Ft. Lauderdale and fully loaded already. Four seats, four people, full fuel and 80 degrees and very close to gross weight. We landed in Pensacola and one of John's friends came up and asked for a ride to Longview for he and his wife, just because they had to get back today.

Being a pilot means you're the pilot in command, and you should know better, but it is also your job. Pilots were a dime a dozen and if I would have said no, I might have been looking for a job in Ft. Lauderdale.

We didn't load up with full fuel, trying to save weight, but it meant we would have to stop in New Orleans for fuel. I knew we were overloaded leaving Ft Lauderdale, but I was afraid to figure it out. We had six people and baggage in a four place Bonanza. I taxied into position in Pensacola and gave the engine full throttle. We slowly started to accelerating down the runway. I had decided early that if we passed the 4000 foot marker and were not airborne, I would abort.

Well, the marker came up and we weren't off the ground yet, but I thought it was about ready. At 4950 feet of a 5000 foot runway we got airborne. We lifted off at 90 MPH, climbing at 90 MPH and climbing up at about 10 feet per minute. At this rate, we would be up to traffic pattern altitude by the time we got to New Orleans.

We finally got up to about 500 feet and I tried to accelerate a little so I could reduce power a little. So I lowered the nose and the speed came up to 91 MPH and we started to descend. Here we were at five hundred feet, 90 MPH, full throttle, and 100 miles to go. We had a problem at New Orleans though.

At the time, New Orleans wouldn't sell gas to light airplanes, but they had a ten thousand foot runway. Lakefront, not far away sold gas but had a short runway. So, I landed at New Orleans, let my passengers out to eat lunch and flew to Lakefront to fuel up, not full though. Then I flew back to New Orleans and picked up my passengers after they ate lunch. Well, it still took over 5000 feet of runway but we made it, all the way to Longview, 500 feet, full throttle, and way over gross.

Another trip, we were going from Galveston to Longview, thunderstorms were forecast for the route, but we figured we could make it. I climbed up to 7,000 under Instrument Flight Rules (IFR) and went on. An IFR flight plan is used when you are allowed to fly in the clouds. We didn't have radar or stormscopes back then. We just went until it got too rough, and then turned around, which was not often.

So we get over Humble. I look down and we're over Humble airport. We entered the bad weather and flew for about an hour. Rough as hell, but John wanted to go home. So we kept on going. After an hour of very rough IFR we broke into the sunshine and I looked down. We were still over Humble airport and the airport looked like a disaster area. A tornado had hit the airport. We had flown through it at 150 miles per hour but didn't go anywhere.

I carried a lot of politicians around the state and around the country, but I set a record with one of the politicians. Every time this particular Railroad Commissioner, Bill Murray, road with me (no matter how rough or how windy) I made a smooth landing. We were on our way to Jackson Hole, WY and it was rough, blowing about 40MPH. Bill was in the front seat as was his custom, and he said, I bet you don't make a smooth landing this time, as said, OK lunch, on landing. He said OK. I came in and the plane was all over the sky as I made my approach.

ZIP, just as smooth as I had ever made.

We went out to the local restaurant and thinking it would be on Bill, I ordered big. We got ready to pay and he didn't have any money. He still owes me for that dinner!

With all of this traveling, John got the bug that he wanted a bigger airplane. Something that would carry more and go faster. I started looking over the airplanes which were on the market. You had DC-3's (too big), Lockheed Vegas(not big enough), twin Beeches, Lockheed Electra's, B-26's, B-25's.

Many of these were big and fast enough but most had radial engines. I decided that radial engines were not cost efficient and wanted something a little classy and odd. We settled on a De Havilland Dove, a twin engine(inverted straight six) cabin airplane with tricycle landing gear.

It would haul six people in the cabin and 2 in the cockpit. The only problem with the Dove was there weren't any in the northern hemisphere, although De Havilland wanted to import them into Canada and the US.

We had to order the Dove from Britain, and they freighted it to the De Havilland company in Canada. The Dove came in pieces. I decided to go to the factory and watch them put it together. Since there were not any maintenance shops in the US familiar with the Dove, I felt I better know everything I could about the plane before we took delivery. It only took two weeks to assemble, but I got to know the test pilots pretty good.

I guess they had confidence in me, because one day one of them asked me if I would like to fly one of their airplanes while I was there. The only planes I saw on the ramp were Vampire Jets the Canadian Air Force was waiting for delivery. I said I didn't know anything about a jet, much less one that weighed 80,000 pounds!

The head test pilot said, "You know how to fly, don't you mate?"

He was serious! I declined, figuring I really wanted to be around to fly the Dove.

Finally the day arrived when the airplane was completed. We started the engines and taxied out to the runway, did the pre-takeoff checklist, and away we went on the maiden flight of what I thought was the best airplane in the world.

The test pilot climbed up to 1,000 feet, rolled it into a 60 degree bank to the right, rolled out, and went into a 60 degree bank to the left and said, "It's all yours mate!"

I packed up my gear, made arrangements to import the airplane, and went through customs at Detroit. Man, this was living.

I didn't have any radios or autopilot in the Dove yet. I thought it would be wise to have them installed in Dallas, where the shop was. So I was going to fly dead reckoning from Detroit to Longview, of course with full fuel. I could have flown 3 times as far. I was really feeling my oats and thinking how lucky I was to fly this kind of airplane.

It had a cabin for the passengers, an air-stair door and it could nonstop across the country. It even had a closet to put my coat in when I got on board. Talk about fancy.

Well, I got on my way, and as was my custom, I lit up my cigarette as I was flying. It is a long way from Detroit to Longview and I ran out of cigarettes in my shirt pocket. Oh yeah, I had more packs in my coat pocket in the closet at the back of the airplane. Uhh. At the back of the airplane. I'm flying solo and am having a nicotine fit, what do I do? Land somewhere and go back thirty feet and then takeoff again? No, I'll just trim the plane slightly nose down, hold the wheel and get out of the pilot's seat.

Then, I run to the back of the plane and retrieve my cigs, no problem. I think running to the back of the airplane was the longest thirty feet I have ever moved in the quickest time. But I get to the back, and low and behold, the plane is flying straight and level. I decide I can move back to the cockpit slowly then and get into the seat and fly away. From then on, one of my preflight checklist items was a carton of cigarettes in the map case behind the copilot's seat! Not a bad adventure in my new airplane on its first flight. I started to believe this could be the best airplane I ever flew.

I finally got the Dove back to Longview and our biggest adventures begin.

One day we were going into Austin and I got cut off by an airliner when I was on final approach. The tower called me up and asked me to execute a missed approach and come close to the tower. Well, the Dove had a clear Plexiglas bubble over the pilot to look up and clear your area. As I flew by the tower I was looking up through the Plexiglas at the man in the tower.

He had such a big grin on his face! He called me up and said we were number one for landing.

Another day we were going into Dallas and the tower asked we could land short and take the first available exit. I said sure, one thing the Dove had was reversible props. It was unheard of in the early fifties. I landed and turned off on the first taxiway. I asked if that was short enough and they said, "uh ha."

After I picked up John I was taxiing out with John up front. I taxied in front of an airliner and the pilot came on the radio and said, " Oooh eeey, that sure is a cute little plane."

The ground controller came on and said, "Yeah, but you ought to see him land!"

I asked John if he wanted to show off a little. He said sure. I made a short field takeoff and kept it climbing at minimum speed. We were at a thousand feet by the time we reached the end of the runway.

The airliner came on and said, "He can climb, too!"

When we went to the West coast of Mexico we normally crossed the border at Nogales, coming and going. The people were friendly there and nearly always we had an easy crossing.

One time for some reason, the American authorities held us up. There was a lot of smuggling going on, and I guess they had heard the rumors as I had.

I was approached by some people and they asked me what I flew. I said a Bonanza. They said if I would go to a certain airport, fly a certain airplane from there to Mexico and leave it for an hour, then fly it back. I would have a brand new Bonanza waiting for me. It was tempting. But I said no.

Back to Nogales, we were going to have to stay overnight, so I left the air-stair down to keep the cabin from getting too hot.

There were no hangars available. The next day, John, Bum, and I went to the airport and there was the dove just as we left it. I always worried about rattlesnakes anytime we were in Arizona. Bum asked if he could ride up front with me and I, of course, said sure. We get cleared for takeoff and get to about a thousand feet. I hear this buzzing sound.

And I think, "Oh, my a rattlesnake is under my seat." I tried to figure out what to do. Do I raise my feet up and try to get out of the seat? Do I just sit here and climb until it goes to sleep and hope it doesn't bite me.

What am I going to do?

About this time, Bum raises up his watch and says, "I hate this damn alarm, it's so hard to set."

I could have broken his watch then and there.

In early 1950, my wife had our second son, Jim, and this caused problems with my job. I took off the last month of Nan's labor so I could be with her on the big day. On the day after Jim was born, John was calling. I did happen to squeeze out an extra day, but nobody in the family was happy with me.

We were off to traveling the world. At least the northern hemisphere. Although we were planning to go to South America. We just never made it. John was always ready for the next adventure, as long as it included fishing or drinking. And, we did party when we were out of town.

We traveled to the west coast of Mexico a lot as I mentioned, but we always seemed to have trouble getting into Mexico. I guess they didn't like the rich gringos coming down. We were in a town on the west coast on one trip. We stayed in a horseshoe shaped hotel on the water. One of the hotel personnel came up to us and asked if we had flown in. We said yes. But we grumbled what was wrong now.

It seems that the governor's son had gotten into an accident and was very critical. They asked us if we could take him to Mexico City to the hospital as soon as possible. Naturally, we said of course and off I flew to Mexico City. I asked the hotel manager to go along since I didn't speak Spanish. We delivered our cargo and the son survived thanks to our graciousness.

On the way back the manager of the hotel asked if we could fly over the hotel. I said sure with a twinkle and a wink. I asked how close did he want to come. He said as close as I want.

I went out over the water and made an approach toward the open end of the horseshoe with the pool in the middle. Did I say I liked to buzz? After we landed and went back to the hotel, they said we made water come out of the pool. Good thing I had the manager of the hotel with me!

A very good thing about taking the governor's son to Mexico City was that the governor owed us now. Anytime we went to Mexico, the border was easy, the hotel, and food was comped, and we had first class service wherever we went.

Later in 1950, we moved to Ft. Lauderdale, Florida, to be closer to the fishing, and further away from John's business. He liked to spend the money, but he didn't want to be bothered with the daily business.

One of the few days we spent in Ft. Lauderdale, I spent at home with Nan and the kids. We were supposed to go to New York the next morning. In the middle of the night, Nan wakes me up and she has had a nightmare. I'm not one to think anyone can see the future or have premonitions.

But Nan had this dream about John and I crashing in the middle of Georgia killing both of us. I was up the rest of the night trying to calm her down. Telling her nothing was going to happen. We would be fine. It was only a dream. This was the first time she ever had a dream about my flying, and I should have thought about it. But, we had to go. I called the weather and they said it should be alright; right there I should have called off the flight!

Nan has been crying all night. I didn't get much sleep, and I was upset. I told her I would call as soon as I got on the ground.

I left her at the fence as I went to the airplane. We left about ten minutes later and got on our way. It seemed like there was a front coming through Georgia that morning and by 10:00 a.m. they expected severe thunderstorms. Just about the time we were expected to be in Georgia. Well, as was John's custom, he went to sleep in the back as I flew. He woke up as we crossed the Georgia and

Florida border. He started telling me about a dream he had the night before, something about going into a front and crashing and everyone died.

I called center and asked for an immediate landing until the front passed through. I landed in South Georgia and called Nan, she was still crying and told her we were on the ground in South Georgia, completely safe. That was the only time I used someone's intuition or dreams as a lead.

John loved to fish and he loved to drive the boat at full speed. We would take the boat through the canals of Florida. The canals had a way of changing position, especially after a storm. We had a friend from East Texas on board and he talked with a very slow east Texas accent. You know, it would take him five minutes to say anything.

Well, we were going through the canals and John was driving. As usual, John was going full throttle and our friend was sitting in the front of the boat watching for channel markers.

Our friend all of a sudden he said, "Mr. John, I see—."

Up on a sand bar we went.

Our friend said, "Mr. John, I was about to say, I see the birds' ankles in the channel!"

A friend of mine, Bud, that I had flown around with a lot got a job flying with La Tourneau. He was a semi-evangelist who started a trade school in Longview. La Tourneau liked to travel as much as John did. La Tourneau didn't want to buy a new airplane. He got a B-26 from the government instead. He refurbished it to make an executive plane out of it. After all, he had a school for airplane mechanics.

When La Tourneau wanted to go, he went whether the weather was bad or not. Bud had told me that he would tell La Tourneau the weather was bad, and La Tourneau would go off in a room and pray. He' come out a few minutes later and say that God told him it was alright to go. If Bud wanted to keep his job, he had to go.

Luckily for everyone, La Tourneau was always right. However, when Bud started his flying job, he was as dark headed as anyone. Three years later, his hair was white as a sheet.

The nice thing about the Dove was we had a lot of room and it would carry a lot. We took off from Longview and John would go to sleep. He would wake up and ask where we were.

I would say, "Half way to New York" and he would say, didn't I tell you, I have to go to Los Angeles first! I would call center and turn around and refigure everything.

We would go from winter in New York to summer in LA. I never knew what to wear. So I carried winter, summer, swimming, tropical, and snow wear wherever we went. Many times we would fly somewhere in the winter and be there for a few days.

The Dove had a liquid anti-icing system on it instead of the pneumatic rubber boots as we have on modern airplanes. However, I liked the anti-icing system better especially when it was forecast to snow.

On final approach to land, I would turn on the liquid and get a good coating on the wings to prepare for the next day's snow. We would leave the Dove outside and when it came time to leave, I would go out and shake the snow covered wings a little bit and "Ouila!" all the snow would slide off the wings. Of course, I still had to climb on the fuselage and get the snow off.

As I mentioned before, John was very generous with the seats in his airplane. I had my family and John's family loaded on board to go back to Longview when someone came up to John. They asked him for a ride to Galveston and naturally he said, "Sure!"

He didn't bother to ask me. I would have told him we were overloaded already with my family and his. We loaded this other family on board and took off for Galveston. No problem.

I never had the situation when I felt like the Dove was overloaded in the air. Even when we had six people sitting on the floor it didn't feel weighted down! I ended up in front of the terminal in Galveston where John lets everyone out. I closed down the cockpit and look out to the left where everyone was walking into the terminal.

We had a solid line of people and babies from the air-stair door to the terminal, good thing the CAA wasn't around! At least when we left we only had two families on board!

John was very generous with my time, and he believed in political contributions. He always believed in giving political candidates a donation whenever he thought he could get something out of it. He would give to the candidate he wanted to win and then he would give one half of that amount to the candidate he didn't want to win, just in case!

I also had to fly around many of the politicians from that era. I took LBJ on many trips when he couldn't afford to go on his own. I even flew the Governor of Texas Allen Shivers during his time. Though, he did not have much confidence in airplanes.

Whenever he and his wife would go traveling, one would go first and then the other would go after the first person arrived. The governor got so confident in our Dove and my ability with it; they would travel together when I flew them.

Things started to go sour with my family since I was gone so much. I had a 12 year old son, a 2 year old son, and an alcoholic wife. Probably from me being gone so much. Anyway, we were halfway from Longview to the Bahamas when I gave John my notice. He might be able to be gone that much, but I couldn't do it anymore.

I told him I would check out the new pilot and do whatever was needed for his safety. John understood, but he never was able to find a pilot that would stay with him longer than a couple of months.

With all that traveling and drinking we did, I guess John turned into an alcoholic. I probably could have very easily. Anyway, before I left, John went to the doctor and had physical. The doctor told John he should quit drinking if he wanted to live much longer, you know doctors, you've got to quit all the good stuff if you want to live longer.

The doctor knew John drank heavily so he didn't want John to quit cold turkey, so he limited John to one drink a day. Little did the doctor know, he should have limited John a little more. John would fill an iced tea glass, 24 ounces, with bourbon and a splash of water about 9:00 a.m. and then about 2:00 p.m he would say it wasn't strong enough and fill it back up. Probably drinking about 40 ounces of bourbon during the day. He never looked or acted drunk.

John ended up living another 18 years. I guess the doctor was right!

When I quit I was unemployed again. Pilots were a dime a dozen in the early fifties. Come to think of it, when were pilots not a dime a dozen? I answered an ad for civilian pilots to teach the basics to military pilots. I think I may have done that before.

Turns out, I was accepted to fly for Texas Aviation Industries out of Hondo, Texas. If you've ever been there, you remember it because they have a billboard at the beginning of town that says, "This is God's Country, don't drive through it like Hell".

I had to go to aviation school again. This time in Selma, Alabama. This lasted six weeks, and it was mostly training in how to teach flying.

Remember, I said we learned to fly during the War and now, in the fifties, we learned how to teach flying. Selma taught us all about spins, aerobatics, stalls, instruments, accidents, and how to teach each of these things. I was one of the few candidates who had experience teaching for the military so it was fairly easy for me. They really didn't want to wash anyone out because experienced instructors were hard to come by.

My roommate was an instructor, but he was also a chiropractor. I guess he wanted some excitement in his life because every night he would go out partying. He would usually settle in about 3:00 a.m. and flop himself on the bed. Every morning, I would have to wake him up and give him a treatment.

He would put himself in some weird position and tell me where to push or poke and snap, snap, he wouldn't even have a hangover. We all partied a lot during this time because we were away from our families and there was not a lot to do in Selma in the early fifties.

Soon, thank God, I don't think I could have partied anymore. We were finished, I went back to Longview, and we started packing. I reported to Hondo for the first class of 1953. I never knew flying could be so much fun!

Of course, when we got to Hondo, the Air Force wanted to know if I could fly the North American T-6. I had flown radial engine aircraft before, but never one as powerful as the T-6. Another problem I had was I had to check out in the backseat. You can't see forward from the back seat.

Taxiing was no problem, you S-turned as you went along to see where you were going. Then it came time for takeoff, I was never so glad to get the tail up so I could see! When you're in the tail low position, you use the sides of the runway to tell if you're straight, but it always feels good when the tail comes up.

The air force check pilot tells me come around and land so I set up the approach, I know I'm lined up straight by the side of the runway, each side should be in the same relative position as the other, but I couldn't tell how high we were. I noticed the check pilot would move his head back when it was time to flair and slow down for landing. I wasn't even looking outside. It was much like an instrument takeoff. I made three takeoffs and landings and the check pilot said, "You made it!"

5 HONDO AND THE KOREAN WAR

We barely had time to find an old farmhouse to live in when I had to report to the airport and start flying. We started the students, who had never flown before, in a North American T-6's, a 600 horsepower, taildraggin (taildraggin means it had a tail wheel), wonderful airplane. The students would get 200 hours in the T-6's, then move on to basic training with bigger and more powerful airplanes.

We would get the students up to point where they would go to fighters training or multi-engine training.

One of the students I had there was named Bill Tower. He wanted to fly fighters from the first time he saw an airplane. The only problem was he was 6'5" and one inch too tall. He knew this but when physicals would come up, he would always try to slump a little. It didn't fool the military doctors. They always said he was 6"5", one inch too tall.

During training, he told me his predicament of being too tall. He tried harder to get good grades, fly the airplane better, and prove that he could fly fighters. I really felt he could fly the airplane well enough and every time he made a turn, he would roll into a 3 "G" turn to prove he was good enough. It came time for his military check ride at the end to tell whether he could go to fighters.

When the airplane came in from the check ride, the check pilot immediately got out and came over to me in the ready room.

"Mr. Atkinson, I have to see you immediately!"

Oh boy, I thought I was in trouble. I knew Bill was a good pilot, did he flunk? The check pilot looked at me and asked in a quivering voice,

"Does Mr. Tower always make his turns at 3 G's?"

I then told the check pilot the reason he did what he did. He wanted to be a fighter pilot!

We recommended Bill be transferred to fighter pilot school, and he made it.

After about a year of too many students crashing the T-6's, it was recommended that we start the students off in Piper Cubs, get them soloed, and then move them into the T-6's. This worked out well, especially among our foreign students.

Occasionally, we would have a student from Holland, England, Israel, and even Saudi Arabia.

One of the few students I had that washed out during training was a Saudi Prince named Sin.

First of all, Sin went through Americanization and militarism at Randolph field in San Antonio. During the time he was there, he was supposed to learn 10 words of English a day. He was there for 6 weeks and should have had a basic understanding of English. He didn't learn a word. He also had a bottomless wallet and didn't like to stay on base.

One of the first things he did was go out and buy a car. He learned to drive in Saudi Arabia where there weren't too many cars, not like San Antonio. He would come to an intersection, get impatient and just pull out. He would also pull out his wallet and buy another car. He crashed 5 cars before the military wouldn't let him off base.

Then they sent him to us. Since I was one of the most experienced instructors, I got him.

Sin was one of those people who liked to do things his way until you got his attention. Since he didn't speak English, I had to come up with different ways to teach him and get his respect.

We started out in the Piper Cub. Since he didn't always understand and didn't want to learn. I would hold his ears and rotate his head to turn or make his head go up and down to make the airplane go up and down. This went on until I could transition him to the T-6. I did get him transitioned into the T-6. He was a decent pilot. He just refused to learn English, and he was stubborn.

The day I picked to solo Sin in the T-6 turned out to be a little strange. I sat down and tutored him on every move he should make in the plane. We started out with the checklist, who he should talk to on the radio, where he should go, and what he should do in the practice area.

Then, we went over his procedure to enter the traffic pattern and which runway to use. He had it all down pat. So, I sent him on his way and proceeded to fly with one of my other students.

Little did I know that the wind would change. Sin practiced in the area as I told him to, and then came back to the airport. He called on the radio and the tower told him to land on runway 32. He set up his traffic pattern for 14, exactly opposite. They had airplanes going head on. They had to do something fast.

I had told him to use 14 and that was what he was going to do. He didn't listen to the tower and the wind was blowing too hard to land on 14. So I get this message from the tower to "Land immediately!"

I came into the practice area and saw what was happening. I got on the radio and followed Sin and talked to him until I could get him to land on 32. After this, even though he was a Saudi Prince, the Air Force decided to wash him out.

There was a problem. The air force check pilot that had to fly with him to wash him out was afraid of Sin. Because of Sin's culture, we couldn't tell him he was stupid or that the other pilots were afraid, because Sin might interpret that as being humiliated.. You know procedure.

The check pilot was afraid Sin might fly them into the ground and kill both of them. So I had to try to explain to Sin what was happening, it was not easy.

Finally, it was agreed that the check pilot would fly the airplane and report that his skill level was not appropriate, but he would be recommended to another flight school that talked Saudi. However the check pilot made a stipulation, he made Sin sit on his hands, then strapped him into the seatbelt and belted him in real tight so he couldn't move his hands. The check pilot started the T-6, taxied out, took off, declared an emergency and landed straight ahead.

I don't think he got 6 inches off the ground before declaring the emergency.

Well, Sin was no longer our problem. I never heard from him again.

To console Sin a little bit, I invited him over to dinner one evening. We didn't know his culture very well. But, we all sat down to eat anyway. As we passed around the food, he would say, "Sol" or salt or something that sounds like salt each time he received food.

My wife, Nan pushed the salt shaker closer to him each time he said it. The salt shaker was almost on his plate when we realized it was a blessing, and he didn't want salt!

We all had practice fields we flew out of to practice touch and goes with the students, just as we did with the PT-19's. One day, I was following a T-6 that clearly had problems; it was up and down, sideways, and just all over. The student would bounce and then go around. I'm sure the instructor was talking to the student and trying to straighten him out.

As we're on the base leg (on the approach to landing), I look out to the other T-6. He was up and down again, then he approaches the ground and bounces and the dust starts flying. And more bounces. And more dust. And finally, all I see is a cloud of dust. Then I see wheels start bouncing out of the dust.

I hear over the radio, "You'll have to call the base and tell them to come get us, it won't taxi!"

Night cross country flights were interesting. The first cross country flight was always a dual ride, meaning it was with an instructor. Then, the students would have to go solo.

Usually, the first one was to Randolph, Texas, then Austin, Texas, and back. The students were not real confident in their ability to navigate at night so we always followed them. More than one would go at a time so it looked sort of like a light line from Hondo to San Antonio. It was really the first student's navigation ability that made it safe for all the students. Of course, being instructors, we wanted to make it interesting. One of the instructors would get in front of the gaggle and slowly turn the line toward Corpus Christi.

Once he had them flying on the wrong heading, he would shut his lights off, and dive away. Then the students would be on their own, on a different heading from what they were supposed to have. Everybody would be talking on the radio, trying to figure out where they were, and eventually they found their way around. After that, we weren't afraid to send them out on their own.

We were required to fly two hours each six months on our own. We would set up a party at an auxiliary field and all meet there. There was only one other instructor at Hondo who had gone through liaison training like I did, so we started having a competition. We would takeoff from the auxiliary field with the fuel tank on the cubs full.

Then we would shut off the engine and glide, it was no problem since there were so many thermals to keep us up. We would see who could fly for our two hours on the least amount of gas. Usually it was only a cup or two gas that was used and no time on the engine. When we would get back, the pencil pushers would look at the hours on the engine, the amount of fuel used, and know that we didn't fly the two hours we were supposed to. Of course, that got us in a bit of trouble. We had to report to our commanding officer and have to explain what happened.

Some people just don't understand competition!

I loved to teach aerobatics to the students, especially unusual attitudes. First we'll talk about instrument unusual attitudes. Then we would practice. When doing unusual attitudes on instruments, the student would close his eyes while the instructor would put the airplane in an unusual or strange attitude and then tell the student to open his eyes and get the airplane to level flight before it stalls or goes beyond the red line in speed.

The way you teach instruments is just like flying while looking outside the cockpit except all the student has to look at is the instrument panel in front of him.

The main instrument to be used is the artificial horizon, which gives you reference to the level ground. Back then, if you got the airplane in the vertical position, the artificial horizon would show a big yellow eye instead of the actual attitude.

First, I would give the students a couple of attitudes that weren't too bad. Then, I would give them a vertical attitude. The airspeed would be going down radically, and this big yellow eye would be staring back at the student. He had about 5 seconds to get the plane back to horizontal before all the air speed would be gone. The big yellow eye would start to roll forward and the student would follow it, sometimes we might reach a negative 5 "G's" as the airplane went over the top.

Then the student would throw up.

Toward the end of the aerobatic training, I would demonstrate an Immelmann with a loop on top. This particular maneuver had to be entered at 185 mph with full power, and just when you get vertical on the loop, you pull full flaps and that's the only way, because you run out of speed.

I always told the students about the speed and the full power, but somehow would forget to tell them about the full flaps. I would tell them to go out and practice that maneuver solo. I was always waiting back at the ready room to tell them I forgot to tell them about the flaps. They would come back, some of them shaking a little, some wet where they shouldn't be, and most just kind of quiet. After that, I knew they could handle anything.

We had a very good maintenance crew. In fact, Adam, the crew leader, lived next door to us. Adam was one of those practical jokers, just to keep things interesting. Everyone brought their lunch and Adam had a way of initiating the new mechanics into the South Texas food.

He always carried habanera chili beans in his lunch bag just in case. Habanera chili beans are like fire on fire, hotter than jalapenos, and they have a heat that stays with you.

I like peppers, but habaneras are too hot. When a new guy would show up, Adam would sit with him, and make friends during lunch. Then after lunch, Adam would pop a couple of beans into his mouth and fake chewing, then he'd pop a couple of more in. He was always careful to keep the beans on the side of his mouth where he wouldn't chew them.

The newbie's curiosity would be up and would ask for one or Adam would offer one. He'd bite into that habaneras and all hell would break lose. He couldn't drink water. It would make it worse. And if all the bread was gone, then he suffered all day. Just a joke.

We also had a guy who just washed the planes, cleaned windows, you know, general stuff. He didn't really know anything about engines. He was curious though, and Adam would let him help. Adam was working on a carburetor one day and asked the wash guy to help.

As Adam opened up the carburetor, the new guy was watching and a nest of bird's eggs fell out. The wash guy saw that and wondered all day long how in the world a bird could build a nest and lay eggs inside the carburetor of a flying airplane.

It seems, Adam had gotten there early and put the nest in the carburetor.

We had a T-6 that was a devil during spins. Instructors would take this T-6 out and show students how to execute spins. After one spin the instructor and the student would come in pale and scared of the airplane. They would red X the airplane's log and send it to the maintenance crew. The maintenance would look it over but couldn't find anything wrong.

The air force would test fly it and sign it off. The next time an instructor was demonstrating spins. One spin was all it took, they would come in to red X the plane. Back to the maintenance shop the plane would go. The Air Force would test fly it, and it would go back to the flight line. This went on for about six months. Finally someone suggested the air force go up with a sandbag in back.

The air force test pilot did this. He came back pale and red X'ed the plane. The maintenance crew went over the T-6 with a fine tooth comb. Come to find out someone, before it came to Hondo, had IRAN (Inspect and Replace As Necessary) the plane and put the horizontal stabilizer on upside down.

I have had many people ask me if I was ever

scared in an airplane. I have to honestly say that once you have done a touch and go at night from the back of a T-6, nothing will scare you!

We did perfected our teaching techniques in the T-6. We had a limited amount of time to teach these kids how to fly up to Air Force standards. There were no more gosports. We had real radios and intercoms. Although, they didn't work very well.

When we first started teaching aerobatics, we would roll the plane upside down and fly along there for a minute. Of course, all the dust and dirt would come flying out of someone's pocket. I always kept the T-6 inverted long enough for everything to settle down, err, up. I told the student he could pick out anything on the canopy as his. I got the rest: money, knives, watches, you name it.

The student was always timid to reach up while we were inverted. He was typically afraid he would fall, but the shoulder harness and seat belt would keep him in the seat.

I would have them raise one hand to the canopy then raise the other hand, the idea would come to them that they weren't going to fall.

In those days, most military airplanes only had about 5 different frequencies to use, they were called channels. We could hear Corpus Christi, Del Rio, and McAllen, if we got high enough. Little did we realize that a super-secret airplane was operating out of Del Rio. We would hear strange calls occasionally from these airplanes.

The first call I remember went like this.

"Flight Service, this is Air Force 12345, I'd like to activate my flight plan."

"Air Force 12345 you didn't put your cruising altitude down, what altitude would you like?"

"Flight service, just put down Flight Level 60+"

"Air Force 12345 we must have an altitude."

"Don't worry about it buddy, there's nobody up here but me!"

That was the first encounter I had with the super-secret Lockheed U-2, the kind that Gary Powers got shut down in over Russia. It is a single-jet engine, high-altitude reconnaissance aircraft that was flown by both the Air Force and the CIA.

The next transmission came a while later and it sounded absurd.

"Flight service, this is Air force 34543."

"Go ahead, Air Force 34543."

"Flight service, 543, I have had a flame out over Ohio."

"34543 would you like to declare an emergency?"

"Negative, flight service, I think I can make it home to Del Rio!"

Ohio to Del Rio gliding, what kind of airplane is that? It took another 9 years before we really understood what the U-2 could do.

One of my other students that stand out was from Holland. His name was Snip. At least, that was the easy way to say it. He came to Hondo at 17 years of age, complete with a college education and could speak 5 languages fluently. Of course, I was only interested in English, and he could speak English very well.

Since he was only 17, I took care of him a little more than the rest; the others were more like 23. He would come over for dinner more often and actually became part of my family. Twenty years later he came to see me. When I lived in Missouri, after doing his tour with the Holland Air Force, he became a duster pilot in the Caribbean. I guess he got tired of the cold weather in Holland.

When it came time to retire the big old T-6. As was the Air Force's custom, they brought in the replacements early so we could get a look at them. We were getting a Beechcraft T-34 and a North American T-28. We would start the students off in the T-34 and the transition them to the T-28.

The T-34 looked a lot like the Bonanza. I was very familiar with that. The T-28 was a monster, a 900 horsepower radial engine on the front with the wings standing at over six feet tall at the ends. You had to lower the flaps to be able to take two steps to get up on the wing. Then, there were two steps into the cockpit.

This airplane was huge compared to anything I had ever flown before. It looked like it would never fly with that big old radial engine on the front. Finally, the Air force brought the rest of the fleet of T-28s into Hondo, about 1956. Other Air Force pilots were supposed to fly the remaining T-6's to Arizona.

The Air Force pilots hadn't been in a T-6

since their training days, probably six or seven years before. We quickly gave them a cockpit check and sent a couple on their way. They never made it to Arizona. They couldn't get them off the ground without ground looping the T-6. These pilots were jet jockeys and hadn't flown anything with torque in years. They forgot about keeping it straight.

The Air Force had all the instructors fly them to Arizona, gee, a vacation.

We had to get used to the T-34 and get a few hours in them before we could take students. I was one of the few instructors who was also an instrument instructor's rating. Most of the instructors had instrument ratings but many needed the instrument instructor's ratings, too. So I used this time to teach many of the instructors this rating. Neil, Carl, Don, and many of these instructors went on to work at the Federal Aviation Agency, after the civilian run Air Force training bases closed, much to my advantage in my later career.

After a few months, we had all gotten used to the new planes and were feeling quite confident in our abilities to fly them. I am in the traffic pattern to land along with about 6 other aircrafts. We were doing touch and goes in the T-28.

We no longer had auxiliary fields since these airplanes had tricycle gears which meant we couldn't afford off-field landings. I was on the downwind leg following the next in line, and Barney, another instructor, comes storming into the traffic pattern right in front of me. We were supposed to enter on a 45 degree angle to the downwind leg and follow whoever was in front of us.

I could tell it was Barney by his voice. I figured he had a bad day or a hot date, so I let it go.

The next time I'm in the traffic pattern, here comes Barney, again. He cuts right in front of me and makes me pull out of the traffic pattern. This means I had to turn right, he was too close to follow, and re-enter the pattern on a 45 degree angle. It was a big waste of time. A few days later, the same thing was about to happen again. I see him coming and hear him on the radio. He cuts me off.

So I increase throttle to catch up to him and say over the radio, "Barney, I can fly without a propeller, but you can't fly without a tail."

He immediately pulled out of the pattern, and I hear over the radio, "Thatta boy Fred. Way to go Fred. Take his tail of Fred."

And that was the last time Barney ever cut anybody off!

One day we were all flying in the practice area and this call came in on the radio.

"All Hondo aircrafts land immediately!"

No ifs, ands, or explanations, just land.

We all rushed in, creating a terrific traffic jam. You think Kennedy or LAX is busy, but it was nothing compared to Hondo that afternoon. We had 100 aircraft all converging on the traffic pattern at once. We couldn't figure out what it could be. Maybe War?

Did somebody drop the bomb? It must be catastrophic to cause us all to rush in like bees to a hive.

After we landed, we were told to go to the ready room and wait. When everyone had come in, the commander came in and told us to strip out of our flight suits. The USAF had just made the command for all personnel to change their flight suits from the new plastic ones we had been issued to the old cotton ones we had used before.

It turned out the plastic flight suits had seriously burned an aircrew-man when a cigarette had come in contact with his flight suit. It went up in flames in fact. We all smoked heavily and any one of us could have ended up that way.

The highlight of flying at Hondo was completing the record the squadron set. All the instructors got together one day and decided to make sure that all the students passed the written and flight portions we were responsible for. It took extra work on our part, but everyone passed. The air force didn't believe it. They knew we had somehow cheated. A class had never been completed before without at least one trainee being bushed.

But we did it!

We followed the progress of that class, and nobody busted all the way through.

In 1957, I stepped off the wing of a T-28 wrong, and I thought I wrenched my back. Turns out I had slipped a disc. I was in pain wherever I went. It got to the point where I couldn't fly anymore because of the vibration and g-forces my body was subjected to.

I lasted six months in pain before I went to the doctor.

He recommended that I have a back operation to relieve the pain, or I could take the easy way out and drink a couple of whiskies in the morning and a couple in the afternoon. I had the operation. All, went well except he cut the nerve to my left leg, and I lost all feeling in that leg.

It was really hard to walk when you can't feel anything. I moved around on a crutch for a while until I could at least keep my balance. I couldn't fly anymore, because I couldn't feel how much rudder I was using.

Texas Aviation Industries (TAI) was good to me and kept my pay going and when I was able, I went to the tower and directed traffic. Since I was a pilot and knew all about the airplanes and the pilots. I set a record for the number of landings and takeoffs at Hondo in an hour.

I knew exactly what the pilots were supposed to do and all the planes were the same, with the same speed, some could land long, some takeoff while others were clearing the runway, it worked like clockwork.

I figured I couldn't fly anymore, and TAI was getting to the end of their contract. I could see the light at the end of the tunnel. I realized I better find something else to do. So my family and I bought a farm in Missouri and moved to the Ozarks!

Two buddies from Longview, Texas, in front of my Franklin Sport.

At 75 years old, I was reunited with the airplane that launched my flying career 55 years prior.

Three generations of Atkinsons stood beside my original airplane. Fred Atkinson Sr, (far left), James Atkinson (center left), J.B. Atkinson (center right), and Fred Atkinson Jr. (far right).

A handful of the pilots I taught to fly during World War II. I am kneeling at the bottom of the picture.

In 1942 - 1944, this was my military flight instructor uniform.

Wendall Tarman, a civilian flight instructor who served with me in Bonham, Texas. He gave my son, Jim, his first airplane ride at Longview when Jim was 9 years old.

I didn't change much between my time being a Civilian Flight Instructor for the Army and a Private Flight Instructor in Longview, Texas.

John and me in front of the Beech Bonanza.

The bonanza never smelled the same after one warm fishing trip to Mexico.

T-28 from Hondo

The first airplane I owned since 1933, at St. Clair Memorial Airport in 1964

St. Clair Memorial Airport in 1983

Our hangar in 1972

The whole family in 1965; Kathy (Fred's Wife), Fred, Jim, Nan, me holding Skip (Fred and Kathy's son) and Jessica (my wife's mother.)

Jim and me, 1970

Merle after he won the spot landing contest with me on left and Jim Evans, mayor of St. Clair

Jim taxiing in the T-34 in 1984

Me and my grandson, JB in 1984

Me and Leroy Hoffman at my 50th anniversary dinner in 1981

Me with the chief of the FAA GADO Office in St. Louis

It's good to relax a little in 1981

6 THE FARM AND FORT LEONARD WOOD

We waited until June of 1958 to move to Missouri so my oldest son, Fred, Jr. could graduate high school. He went to the Hondo School District since he was in the sixth grade. Luckily for us, when I worked for John, he allowed me to buy some oil leases in East Texas so we had about $500.00 a month income. That was not bad in 1958 for not doing anything.

We got to Missouri and remodeled the house, got Fred Jr. set up in college, Jim set up in elementary school, and I got to farm. The farm was fairly hilly but I turned it into a sheep farm, in an area where everybody raised angus cows and milk cows.

I always did things a little different.

Anyway, about a year later the oil leases we had were declared slant wells, and we lost all that income. The farm had taken all the oil income to sustain itself and provide money for us.

During this time, I decided to apply to the

newly formed FAA as an inspector, pilot, or whatever they'd be willing to hire me for. While I was at Hondo, I gave flight instruction to a lot of the instructors who later went on to establish the modern FAA.

I really didn't think I would have a hard time getting in. My application came back stating that I didn't have enough education. Stupid idiots. I know how to fly! And I taught them to fly! In fact since then, I have taught many people to fly who went on to the FAA, airlines, and military.

We went on for another year, but I had to find some work. There just wasn't enough money to keep us going. A friend of mine had a son, Harold who worked at Fort Leonard Wood as a pilot and instructor for the flying club.

He said business was booming, and they needed pilots for their Cessna 195's and Lockheed 10's. They also had a Piper Tripacer and a cub to teach in. It was sixty miles away, but the money sounded good for 1961. It paid $4.00 an hour. So I went over to meet Aubrey, the owner of the charter service and flight school. I only got paid for the time I was in the air so I volunteered for instructing, charter, whatever.

The only thing I worried about was my dead leg. Could I fly again? Nothing said I couldn't as far as a physical goes, so I figured I might as well try. It took a while, and I only went up with advanced students for about a week, but I got my technique down and nobody could tell I had a dead foot.

The planes we flew always had oil leaking out, of course, radials always leak, but not like these. Oil was always down the side and sometimes covered up the windshield. I don't know for sure if they were inspected or not. We flew anyway. I was used to Air Force maintained and rich oil man maintained aircraft, not these hunks of junk.

However, I must say, they never quit on me, and I always got where I was going.

The Lockheed 10 I flew was notorious for losing its brakes and it had a free swivel tail wheel which meant I had to use the engines and brakes for directional control on the ground. I got in the habit of doing a brake check on final and more often than not, I wouldn't have brakes.

One night I was going into Memphis with a load of soldiers on a weekend pass in the dead of winter. The runway had an inch of ice all over it from a rain early in the day. Clearing runways was not the best. I approached anyway. I checked the brakes and nothing. The only runway had a 20 MPH wind down the runway, and I made a very nice landing. I got the tail down and was maintaining a pretty straight path and I slowed down and got ready to turn off the runway.

I had no brakes, icy runway, lots of wind, and I couldn't turn. The airplane only wanted to stay pointed into the wind. I called ground control and told them I couldn't taxi, could they send out a bus for my passengers? We had to tow the Lockheed to the hangar, and I got the brakes fixed before we left.

Every time I went into Memphis after that, the tower always asked if I needed a tow truck!

The reason we went to Memphis and

Chicago all the time was that the soldiers on a weekend pass could only travel so far. I think half the soldiers at Ft. Wood were black and from Memphis and Chicago. I was coming back from Chicago one night, looked back in the cabin and all I could see were teeth and eyes!

Another evening we were going through some bad weather with a lot of lightning around us. The Lockheed had a couple of antennas on the nose of the plane, pointing at the windshield. One of the passengers asked if he could ride in the copilot's seat since it was empty. I said sure, he sat down and all of a sudden the lightening got closer and worse.

We had St. Elmo's fire coming off the antennas and splashing on the wind shield. The fella slowly got up and moved all the way to the back of the plane.

Another time going into Chicago was really rough and the plane was full, including the copilot's seat. It was cold, about 10 degrees outside and most everybody in the plane had used a sic sac one time or another, except the guy in the copilot's seat. I was making an Instrument Landing System (ILS) approach to Midway. An ILS is a guiding line to the landing strip. We couldn't go to O'hare, it was too far, and all of a sudden the guy next to me decided to spew, all over the instrument panel!

I couldn't see the instruments. Luckily, I was on the ILS when it happened and just continued my approach. When I broke out I opened the side window for air and to see. I had puke all over me, the instrument panel, floor and cushions. I had to go to the hotel to clean up. By the time I got back, it was all frozen and harder to clean than it would have been.

From then on, I made sure the passenger in the copilot's seat had a sicsac!

Harold, the guy that got me hired, didn't have a twin engine rating and he asked me if I could help him out. It had to do with flying so I said sure. A couple of times he would go on charters with me and get his time in. Once in a while, we would deadhead back so I would be able to teach him a little more about single engine flying in a twin.

A deadhead is traveling without passengers or freight.

We were at 12,000 feet and I cut an engine. He was beginning the checklist and feathering the bad engine's prop. He made a mistake and feathered the other engine. We were at 12,000 feet when we became a glider.

Unfortunately, once you started feathering the engines on a Lockheed, you had to complete the process. It took about 20 seconds for the good engine to feather and then another 20 seconds to come out of feather.

After we lost 10,000 feet because of drag and no engine power we finally started flying again. We didn't practice stuff like that with passengers on board.

After we would get to altitude, the passengers would start walking around and get a little familiar with the airplane. One of them would walk up and ask questions like, "How high is we?" "How fast is we?"

To say the least, these guys weren't world travelers.

I would have charters for the government carrying prisoners to or from Ft. Leonard Wood. Most of the time, there would be two guards and the prisoner in the Cessna 195. One particular time, the guards brought the handcuffed prisoner to the airplane and handcuffed his hands to the framework and then handcuffed his ankles to the seat framework. I told them I wouldn't be responsible for the prisoner, and they said I wouldn't want him loose!

The guards told me their weapons were loaded too and I believed them.

This was in 1962 and General Curtis LeMay was commanding the Air Force. They were very particular about Strategic Air Combat (SAC) bases. SAC bases were a base for bombers who carried nuclear weapons in the '50s and '60s. Whenever I went into military bases I always filed a flight plan. I had to deliver a prisoner to Little Rock AFB, a SAC base, and I called in for clearance to land.

The tower came back. They claimed the approach and landing was unauthorized. I wasn't about to take this prisoner back. I probably wouldn't get paid. So I tried again and told them I had filed a flight plan and had a prisoner on board for delivery.

Again, "approach and landing unauthorized".

I talked to them for about 15 minutes, who was on board, where we were from, our secret clearance, pilot's license, mother's maiden name, etc.

Finally, they cleared me to land and said, "Turn off at the 12,000 turnoff only."

I usually land short and this time I didn't touchdown until I had 4000 feet left and turned off at the 12,000 as instructed. There was a "Follow Me" Jeep with two Military Police's in the back with their M-1's pointed straight at me!

Not the other guys, just me! Well, I followed them for about a mile to the guard shack and those two guys were still pointing their guns at me! It seems someone had lost my flight plan and we were intruders.

After a few phone calls, it all got straightened out. Except, they kept those two guns and their M-1'S on me!

We got our clearance to leave, went to the airplane, followed the jeep to the intersection for takeoff, and there were those two MP's with their guns on me! I was never so glad to get out of there. I had to go back a couple of times after that but the flight plan always preceded me.

There was a restricted area over by Forney Field, where we flew out of. I never really bothered staying out of it and many times would cut across it on my way into the pattern. Then I was flying one night and getting ready to cut across the area to enter the traffic pattern. I saw tracer bullets shooting up into the sky, right where I was going. The bullets were being shot at targets and deflecting off the mound of dirt behind the targets and going into the sky. That was the last time I even got close to the restricted area.

After I was at Ft. Wood for a little over a year, Aubrey kind of got a big head. He was running a small airline, or so he thought. All we had were two Lockheed ten's and two Cessna 195's and three pilots. The planes were always filthy, oil streaked, and in need of repair. But Aubrey thought we should all be wearing uniforms. I flat out refused. He flat out fired me.

Here I go again. A pilot out of a job. Out of money. And nothing on the horizon. I couldn't wait to get home and tell my wife. It was not a good experience when I got home.

7 VICHY

I got home and all hell broke loose.
She said," How can you quit your job?'
I said I didn't want to wear a uniform, and it went downhill from there.

As much as I would like, there were not many flying jobs in the Ozarks at that time. First thing I did was go to Rolla National Airport at Vichy and see Lee. Lee was the airport manager and flight school operator. Lee didn't really like instructing, but I did. He did like to take charters and fly fire patrol which I didn't like. So we kind of fit together.

I would drive about the same amount of miles I had to Ft. Leonard Wood so that wasn't too bad.

I got home and told Nan I had another flying job. At least I wasn't sleeping on the couch. I went 5 days a week to Vichy, as I had to the Fort. That didn't leave very much time for the farm so it went downhill pretty fast.

After about six months at Vichy, I decided to move the family to Vichy. It would help on mileage, and we could rent out the farm. We rented a house on the airport for the outstanding sum of $25.00 a month, utilities included.

Of course, I was only making $4.00 an hour flying.

Bob L. owned a Cessna 140, and he had a private license. He worked for the conservation service. They told him if he got an instrument rating he could fly for them. So, Bob was one of my first students. The 140 had a Ventura Tube instead of a vacuum pump so it took ten minutes of running at high rpm to wind up the gyros for the instruments to work.

Most of the time all we had was partial panel. We filed our cross country flight plan with the Vichy flight service. We were going from Vichy to Springfield, Missouri, and it turned out there was a 40 mile an hour headwind so our ground speed was in the neighborhood of 50 mph. This was 1963, well before radar coverage was over the entire US.

We made our first position report and according to our figures. It would be about 40 minutes until we got to the next reporting point.

About 10 minutes later, center calls up and asks if we're at the next reporting point.

No, we said.

About 5 minutes later they called again, asking if we had reached the reporting point.

No, we said.

Then, they asked if we had an estimate for getting there. We said, "just a minute."

Bob sat there and figured and he reported it would be 25 minutes.

Center comes back with, "are you lost?"

No, we replied.

"What is your groundspeed?"

"50 mph."

"What kind of airplane are you?"

"Cessna 140."

"Oh, we thought you were a military C-140." Big difference!

Lee had a 1962 Cessna 150 for training. It was tiny compared with newer 150's. It did not have a back window, but it did have tiny wheels.

I had a student named Doc. He was wide in the hips and narrow at the shoulders and I was narrow at the hips and wide at the shoulders. We fit in that old 150 like two old hourglasses.

I weighed about 165 lbs and he came in well over 300 lbs. The Cessna 150 was over gross every time we flew it! Well, Doc chewed cigars. I don't remember ever seeing him smoke one, but he always had one in his mouth. The more nervous he got, the more that cigar moved around his mouth. When he had finished the end he was chewing on, he'd turn it around!

The only way we ever were able to get in the Cessna 150 was for each of us to open our doors all the way, yell, "exhale," and slam the doors. I got him soloed after about 15 hours. I think he liked me. I would tell him that he was going to solo the next time, and then he wouldn't show up for a couple of weeks.

He'd come back a little rusty. I worked him through it again. Then I'd tell him he was going to solo next time, and he wouldn't show up. He could land just fine. He was just scared. So, the next time he showed up, we did a few touch and goes and finally I told him to stop in the middle of the runway. I got out and told him to takeoff from where we were.

It was a 5,000 foot runway so he had plenty of room. I just stood out there beside the runway and waited for him to come around. He made a real sweet landing and he taxied in. He forgot all about me on the runway. I had to walk in. Or, did he forget?

One day, Doc was riding along in the Cessna 150 with another pilot. He wanted to navigate. Luckily, the other pilot was a small woman named Valera. I watched them takeoff, and just as the Cessna 150 lifted off Doc opened the sectional all the way across the cockpit. Talk about an instrument takeoff! All Valera could see was the sectional. Luckily they got airborne before any other problems arose.

After about a year, I had students driving 60 miles just to fly and learn from me. The FAA had heard of me, and they came down to fly check rides with my students. One of the examiners who came down was Pete Campbell, a well known FAA man and for a while in the '80s he worked for Airplane Owners and Pilots Association (AOPA.)

Pete could ride with someone and even though the student tried hard. Pete could tell that student with a smile, "son, you need a little more time."

When I became an examiner, I never could do that. I always felt sorry for the student, they usually try so hard. So one day it came to pass that Pete came to me and told me, didn't ask, told me. "Fred, we need an examiner down in this area and you're it!"

How could I say no?

I became an examiner for the Private and Commercial Licenses in 1963. I remained an examiner until 1983 when my son, Jim, took over for our area.

One of my students that Pete flew with was Dinger. Dinger had three pair of glasses to wear, one for reading, one for instruments, and one for driving. I guess he didn't need one for walking. Well, the day came for Dinger to take his check ride. He forgot his instrument glasses at home, 60 miles away, so he decided to take the ride anyway. According to Pete, everything was fine except his instruments. It was like Dinger couldn't see and was all over the sky!

So he flunked the ride. The next time he passed was when he remembered his glasses.

My oldest son, Fred, Jr. was attending A&P mechanics school at Parks College. His roommate for one semester was Jim D. who happened to want to fly airplanes more than work on them.

When they graduated from Park's Jim's

ambition was to fly for the airlines. So Jim's plan was to get his commercial license and apply to the airlines. Somehow, he got talked into learning to fly at Vichy from me. Jim moved right into our house and proceeded to learn to fly. He never went out and always studied. I was getting a little worried about the kind of person I let into my house. I couldn't talk him out of the airlines. He was so determined. He went right through the private license, but decided not to take the check ride. He didn't want to take anybody with him. He just wanted the commercial.

Jim eventually had to take the Private check-ride before a trip.

One of my other students bought a Cessna 195, but he didn't have a license yet. He decided he needed to fly to Texas for a meeting in a couple of days. I hadn't yet checked him out in his toy; we had only flown a Cessna 150. But he wanted to go; I couldn't go with him because of prior commitments so I worked out a plan.

Not quite legal, but it worked. Jim had gotten his Private, but he didn't have any tail-wheel time or radial engine time. Lee's son Gary, had 100's of hours in a Cessna 195 but he didn't have a license yet, he couldn't pass the written.

So, I got Gary to fly the airplane, Jim to sit in the right seat, and the owner sat in back! They made the trip, no problem.

I had a dubious honor at Vichy. You see, we had a GI school to teach the ROTC students at The Missouri School of Mines at Rolla. I would teach about 5 students each semester and within the allotted time of 35 hours our students would receive their Private License as the Army wanted.

It seems that most of the other college schools could not get their students to graduate in the allotted time. The Army came down and inspected our outfit. They flew with the students and were very satisfied with the progress of the students.

The other schools teaching the college students were reprimanded and their training was cut down until they could prove they could do what we could.

Since I was teaching some of the students at the school, some of the professors decided to learn to fly also. One of them went on to become an FAA inspector. He was teaching instructors how to evaluate other instructors to become flight examiners.

Another of the professors did not have the same drive. He would come to the lesson and after everything I said he would say, "yes, yes, yes," as if he understood.

He didn't understand and thought he was God's gift to humanity. Finally after a few hours of this I told him to stop trying to fly the airplane and listen to me. Again, he said, "yes, yes, yes."

I said, "Doctor such and such, you're one of the best instructors at the school of mines aren't you?"

He kind of beamed and said, "yes."

I looked at him and said in a very firm voice, "I am the best instructor around here and what would you do if one of your students treated you as you are treating me?"

Well, he kind of calmed down after that until I got him soloed. Thank goodness he quit. I think I would have quit flying if I had to keep putting up with him.

Occasionally, I would have to take a charter for Lee or fly fire patrol. I had a trip down by Camdenton and ended up having engine problems. Luckily, I landed at the Camdenton Airport. Oil was dripping everywhere, and I didn't know what was wrong. I called Lee and told him about the trouble.

Lee said, "Just go ahead and fly it back, it'll be O.K."

"Hell no, you fly it back."

Lee came back and said, "If you won't. I sure as hell won't."

Valera was one of the pilots I trained, and she liked to fly through thunderstorms. Whenever thunderstorms were forecast she would come out, and we would fly around looking for some to fly through. I was just crazy enough to enjoy the ride also.

One of the last times we did this, the storm was not particularly severe. The airspeed in our Cessna 182 would go from 70 to 180 in the blink of an eye. The vertical speed would peg itself from climb to descent. It was a carnival ride.

We came back and landed, when we exited the plane we inspected it and found a few rivets had been pulled out of the wing and tail. I guess that's a good reason not to fly through thunderstorms and always fly around them.

Speaking of carnival rides, with all the instructing and aerobatics I do, I still can't watch a merry-go-round or a Ferris wheel. I'll get sick.

Valera was a good friend. She always brought me students here at Vichy. And, as fate would have it she told me they were building a new airport at St. Clair, where she was from. They needed an Fixed Base Operation (FBO) and airport manager. I thought, why not?

8 CIVILIAN FLIGHT INSTRUCTION

I hurried up, sold the farm for a loss, and borrowed money from the bank in St. Clair. After that I was never out of debt for 22 years. To say that the banker was skeptical of my business venture would be optimistic. The banker decided they could take a chance on me, because the city needed someone to operate the airport.

After all, I was only asking for a $10,000 loan, and I was buying an $8,500 airplane with that. It sounded like a good deal to me.

I started business at 8:00 a.m. on August 4, 1964 with my first student, Lee J. E., in the second airplane I had ever owned. It was a brand new Cessna 150 number N6004T. He soloed after about 8 hours. Not in the same day of course!

Our office was a tree at the back of the property. If people wanted to fly and I wasn't there, they would leave notes on the tree. A friend of mine had a Cessna 170. We used his radio as Unicom base, because we could taxi his airplane right up to the tree.

One of the pilots at St. Clair that I didn't teach, went out and bought a Cessna 170. It was a tail wheel type airplane. He wanted some dual to learn wheel landings, where you land on the main gear with the tail up. In some airplanes it is quite tricky. He asked me how to wheel land, and I demonstrated the technique to him and greased one on.

I said, "O.K. it's your turn."

He came around the pattern and greased one on just as I had.

I said, "do it again," knowing this was just an accident. Leroy was very full of himself as he came around, because he thought he was just as good as me. He set up his approach and hit the ground and bounced higher than a hangar. Here we were, thirty feet in the air, no airspeed, and ready to fall back to the ground.

He started to slam full power in to recover. In my mind, I saw us roll inverted because of the torque. I just let him ease in the power and nose down to start accelerating. We went around and never touched the ground. Whew!

After that, he made many wheel landings but never bounced as high!

I have had people come up to me and practically demand to fly for over an hour each session. I suppose they want to get their license quickly, but at the end of an hour I can see their physical and mental abilities diminish rapidly and the more they fly the worse they get.

This philosophy works for non-commercial flyers, and those pilots who don't fly often. It seems that pilots who receive more total time increase their ability to assimilate information better over long periods of time.

I suppose there are some instructors out in the flying community who would fly 2 or 3 hours in one sitting with a student and expect the student to learn. Of course, the instructor may not have been teaching very much anyway. I teach and give out so much information in an hour of flying time I spend with a student that I find the student needs time to assimilate the information.

Many times I have found the student's subconscious mind absorbs most of the information. I use my own form of psychology to teach people to fly, and I only resorted to physical abuse once. When I am teaching, I always had to talk loud enough for the student to hear me, over the sound of the engine. Sometimes, the student thought I was yelling at him, but most of the time it was merely trying to make sure that I was heard. We didn't have headsets at that time.

Harry, one of my students, had a problem called fixation. He would be approaching the end of the runway and stare at the near end. He would raise the nose to try and get to the end of the runway instead of adjusting his power to get there. This would slow the airplane down and increase his descent rate. Then, he would notice he was going to be short and raise the nose again, compounding the problem. We call this reaction, stretching the glide.

After Harry had tried to stretch his glide for about 4 hours I was getting a little perturbed. He wouldn't listen to me. And his uncontrolled gliding could cause him to crash.

I very seldom got mad in an airplane, but he was doing it to me. I was so mad at Harry. I took my right fist and hauled back. You can't do that very well in a Cessna 150. And I hit him in the shoulder. Then I hit him again. And again until he quit stretching his glide.

I hit him so hard I popped the door hinge on his door and hurt my wrist. Harry went on to get his license. He even owned a Cessna 172, but he never stretched his glide again.

The four fundamentals of teaching are explanation, demonstration, practice, and critique. When I taught I would never tend to show impatience until the practice time came. I tend to become quite aggressive during the practice time. When the time is right I want the student to feel anger. I feel that anger and fear invoke the same feelings in the body as adrenalin and behavior. I do this because I want to see how the student reacts to this stress.

I never want the student to feel fear in the airplane, but I do want to see how he'll react to the fearful situation. So, I make the student mad, maybe at me, at the airplane, at himself. It doesn't matter as long as they are mad. I want to see the reactions of the student and make sure he makes the right choices while under duress.

If I am not satisfied with their choices, then later I lead them through stress again and show them how to make the right choices. Then before we get back to the airport, I make sure the student sees that the anger he was feeling was brought on by me and he understands what happened, otherwise he might not come back!

I feel the student should be put in as many adverse conditions as practical so if he encounters a particularly troublesome problem he should be able to handle the problem. I like to take the students out in strong crosswinds (over 20 knots), low visibility, and rough air. And then if the student encounters these problems solo, he can handle them.

I don't send students out in those conditions solo, but many times the weather changes in the middle of a flight.

One day I had a charter flight and had to come back and land at St. Clair. As luck would have it, the wind came up above 40 knots in a direct crosswind. The FAA dropped in for a surprise inspection just before I landed. My wife was in the office with the FAA, and they heard me call on the radio as entering the traffic pattern.

They were astounded, how could someone land with a forty knot crosswind. I came in and landed in a sideslip, left wheel first and stayed on that wheel for about 100 yards then leveled the airplane and taxied in.

After I let my people out this FAA man came up to me. I could tell he was mad.

He said, "Don't you know you can't land in wind like this?"

"Well, I did, didn't I?"

He was new. Eventually we became friends, but he sure was quiet after that.

We had a strange situation occur on a flight from St. Clair to Vichy. We had several cats around the airport for rodent control. Bill H. decided to fly to Vichy and he pre-flighted the airplane, got in, and took off. Everything was smooth and clear.

All of a sudden he hears a screech and a cat grabbed the back of his neck with all four sets of claws. Bill about jumped out of the airplane. After losing a lot of blood, Bill managed to remove the cat from his neck and found a box to put the cat in. The cat was relatively calm for the rest of the flight.

We were one cat less at the airport when Bill got back, and he always checked the backseat after that.

I had a dubious honor in the sixties; a couple of other airport operators sent their sons to me to learn advanced licenses, instrument, and instructor' ratings. Now granted, when you have a son you want them to learn in the best way possible, but if these other flight schools were good enough to charge other people to learn to fly, then how could they in good conscience send their kids to me?

I taught my own sons to fly. I have made them both very angry at me. I suppose I expected more out of them than other people. It was not an easy experience for them or me. I made my oldest son so mad at me during some instrument training that he ripped the instrument hood off his head and started for home. I slammed the hood back on his head and said, "I want you to go to my funeral. I don't want to go to yours!"

It took him a couple of weeks to get over that session. When I was teaching my youngest, Jim, he was still living at home. He would go days at a time without speaking to me. He was young and had to let his age catch up with the FAA age requirements. He soloed on his sixteenth birthday in a freezing rain. Private on his seventeenth. Commercial on his eighteenth. His commercial check ride was the best I ever had. He could really do chandelles and lazy 8's.

I finally got Harry T., another student of mine, to the point of soloing, and I told him he would solo the following lesson. Well, Harry was quite a character around town; everybody knew him and liked him even though his language was real rough. Every other word was a cuss word, but it just sounded natural coming out of his mouth.

Harry told his wife he was going to solo. He shouldn't have done that. She calls me and tells me she wants to have everyone come out and watch his solo. I tell her not to come out until 4:00 p.m., roughly the time he's going to solo. I didn't want them out there when he comes out. It might make him nervous.

So, everyone is hiding when I get out of the

airplane and Harry takes off. Just for fun, I get on the radio and tell Harry everyone in town was here to watch him solo. He comes back over the radio and says," I don't care how many ****ing people are there, ****, *****," and a few other choice words, then he realizes he's on the radio and not supposed to cuss. He gets half way through "sons of bitch" and shuts up completely. He did make a good landing though.

    I was performing many check rides during this period, but one in particular stands out. My next check ride was flying into our 2,500 foot airport for a commercial check ride in a Cessna 172. The pilot set up his approach to land on runway 2, slightly downhill but into the wind. He decided he was too high. He wouldn't be able to land and stop before running out of runway. So, he made a missed approach and came around to land again. I thought to myself, he's using good judgment. Remember, I'm on the ground and haven't met him yet. Each time he set up his approach, the base leg was closer to the runway than the last one and he was higher each time. Finally, after five attempts he put the airplane on the ground and landed so fast he nearly ran out of runway.

    He taxied in, shut down, and introduced himself as my commercial check ride applicant.

    I said," I think you should go home and get some more dual on landings."

    "Why?"

    "As a commercial pilot you should be able to land an airplane properly, not as you did. And, you should make the necessary corrections to do things right the next time."

"But this is the first time I ever landed here."

"All the more reason to plan your approach and make sure you put the airplane down on the ground in a smooth and easy manner."

"Well, would you fly with me until I'm good enough?"

"No."

I never would, nor, did fly with someone before a check ride. Even with my own students. I would let another instructor recommend and fly with them for the last few hours. I never had a disproportionate number of applicants who did not pass, but many times after I failed someone they would come up to me years later and say, "I'm sure you don't remember me, but you failed me on a check ride ten years ago. I just wanted to thank you. I learned so much more after the failure and because of it. I found out flying is not a game. It's life and death."

Of course, there are others who feel they deserve to pass the check ride just because they have the minimum amount of time. The minimum time is just that, minimum! Most of my students took their check ride when they had about 48 hours, the FAA minimum is 40.

I found that the minimum was just not enough time to teach the bare essentials. I was never one to accept the minimum requirements for my own students. If the book said the minimum altitude deviation in the traffic pattern was 100 feet plus or minus, I tried to teach my students to fly within 50 feet.

Students or instructors would call on the phone and sometimes it went like this:

"Airport, Atkinson."

"Hi Mr. Atkinson. This is John Doe. I am a student of Avery Bush in Podunk. I would like an appointment for a check ride for the Private License."

"OK, when would you like? I am available most days."

"Well, I have a couple of hours off from work next week and I wondered if we could do it Monday afternoon. I can be there around 3:00PM but I have to be back to work by 5:30."

"I'm sorry, but the check ride will last about 4 to 5 hours. You won't be flying that long, but with the oral test, preflight, flight and paperwork, it usually takes 5 hours."

"Well, Joe Blow will do the check ride in an hour."

I'd replied, "I guess you'd better go to him."

That would usually discourage him enough to go to the other guy. I always wondered how someone could give a check ride in one hour. I even heard of a Private check ride that lasted 30 minutes. How can someone adequately judge the applicants ability and skill in thirty minutes?

At least nowadays the FAA is trying to correct some of the problems in the system. The Practical Test Standards are a great improvement of the flight test guides. Theoretically, with the old flight test guides, if the student didn't scare you,

you could pass him. The Practical Test Standards give the instructors and the designated examiners parameters to stay within. Some people will not follow it anyway. But perhaps the FAA can weed out those mustangs.

Looking back over the past check rides, I haven't changed my check rides except for minor changes in 21 years. I followed the Practical Test Standards long before they came out.

Jackie and Felix came to me one day and they wanted to learn how to fly. They would always come to the airport together and one would wait on the other one. They were practically inseparable. I taught different people different ways and Felix was very mechanical and Jackie was methodical. I had to come up with the right way to teach each one of them. Well, after about five lessons they started comparing notes about what they were learning.

Naturally, I was teaching each one a little differently. One big difference was the language I used with each of them. Jackie confronts me and says she wants to learn just like Felix. She could take the language. To make my point with Felix, I would use a lot of cuss words to get my point across. When Jackie asked to be treated just like Felix, I did what I was told.

We went up and I started chewing and cussing. She seemed to do alright, so I kept it up for an hour. We got out of the airplane and she was real quiet. She sat down in one of the office chairs. Then, Felix and I went out and flew for an hour. We get back and she's still sitting there. We ask her if she is alright.

She said, "I've changed my mind. I don't want to be treated like the men!"

They eventually went all the way through and became very good pilots. They bought an Ercoupe and flew all over the country in it. Then they made a deal to buy a newer ercoupe. Jackie wanted me to fly with her in the new one and give her some pointers. We were taking off to the south over town and all of a sudden the plane started shaking all over. I couldn't even see the instrument panel. I took over and cut the power but that didn't do any good. We could kind of control the plane, but it still shook like hell.

We were about 400 feet high when this started, and there wasn't much wind so I did the thing you're never supposed to do. I turned back to the runway. We kind of glided back. I tried to turn the engine off, but that didn't work. I couldn't figure out what happened. We are gliding back to the runway and landed without any problems.

We get out of the airplane and start looking over the airplane. It appears to be alright. We work our way to the engine. There was 7 ½ inches of the propeller missing! Felix ordered a new propeller. He investigated the history of the airplane and found out it had a propeller strike and damage to the front of the airplane just before they bought it.

We didn't have anything like car facts on the airplanes we bought at that time.

We've had many amazing and sometimes funny things happen at St. Clair. One day we heard a call to Sullivan Unicom from a King Air, a town about 18 miles west of St. Clair.

"Sullivan Unicom, this is King Air 38L Airport Advisory."

We didn't think anything more about it except about 5 minutes later we had a King Air land at St. Clair. We went out and greeted the passengers and pilot as was our custom for all airplanes.

Five executive type people deplaned and then the pilot.

The pilot asked, "Has a car been here to pick up my passengers?"

I replied, "No, nobody's been here all day."

I asked him where his passengers were going and he said some factory I had never heard of.

"I'm sure they'll be here in a minute."

Boy, was I wrong!

"I notice you have a fairly new airport, was the old one on the southwest side of town?"

"No, the old one was right across the highway, but it closed 10 years ago.

"Well I have a map which shows it on the southwest side of town."

"Could I see that map?"

I was curious about oddities and knew I needed to send in a change on the map.

He showed me the map and sure enough, the airport was on the southwest side of town, in Sullivan, not St. Clair. The King Air pilot only missed his destination by 18 miles. He loaded everyone back up into the King Air and flew the 18 miles to Sullivan.

Was that considered a near miss?

Speaking of near misses, how can two aircrafts run into each other when both pilots are supposed to be looking outside of the plane?

I was in the traffic pattern one day and looked out the side window when I noticed an F-4 approaching our location. There wasn't anything I could do. He was moving at 400 mph and would be on us before I could move the Cessna 150.

As he approached, he rolled into a vertical bank and turned on his afterburner, he passed within 20 yards of us. He was so close I could read the words on the belly if I wasn't so shocked.

If he would have turned on his afterburner first, and then rolled in to the bank, we would have been toast.

In this case, we were both looking out the window, but he was approaching so fast, we couldn't see each other until it was too late. I thank that F-4 pilot because he knew what he was doing.

In the late 60's there were many new style aircraft coming out, and one of them was the Mitsubishi MU-2. An obvious Japanese fellow called into Lambert while I was listening and said, "Lambert tower. This is Mitsubishi 23T," in a thick Japanese accent.

The tower comes back and says, "23T, say aircraft type again."

"23T is a Mitsubishi."

"23T say aircraft type slowly, one more time."

Someone else chimes in with "Turbo Honda!"

My son Fred was manning the airport when a pilot in a Cessna 140 comes into the office. He asked if we had a new airport. No, it's been here for about four years. Here we go again. He had a chart showing the St. Clair Airport to be across the road.

Fred said, "Let me see that chart." Sure enough it showed the airport on the north side of the road. The old airport that had been there in the forties. Fred thought quickly and told the pilot that was a misprint but here is a new one, and we better get rid of the old one. That was the only chart I have ever seen with the old airport on it, and I still have it.

Harold had just lost his wife and needed a hobby. He heard about flying from some of his friends and decided to take it up. He was listless and didn't really try too hard. I couldn't push him, knowing what shape he was in. So, I didn't do the usual fear and shake 'em up as I did with the other students.

I got him soloed after 14 hours and he seemed to liven up a little but not much. He got about 20 hours when he was in the traffic pattern shooting landings and takeoffs just as I told him to.

My son Jim was watching him to make sure everything went alright, since I was with another student.

Jim said that Harold made a little shaky landing and then the flaps came all the way down and Harold was heading for the highway, about 600 yards away and not down the runway. He was airborne, but only in the ground effect.

If he would have kept going and milked the flaps off, he probably would have made it, but he decided to land in the open area between the runway and the highway. He touched down and promptly turned over.

By the time Jim had run over to the plane, Harold had loosened his seat belt and fell on his head. The only injury he had was a scratch on his head from loosening the seat belt. After that, Harold was a different man. He seemed to enjoy life, got a girlfriend, finished his pilot's license, and bought a Cessna 172.

That accident changed his life, and I wouldn't have it any other way. I did blame myself for the accident though, because I didn't push him.

Jim P. was one of the airport bums who used to fly in the air corps. He flew B-24's in the Pacific Theatre. When he got out of the air corps, he felt like he never wanted to see an airplane again. I'm sure being shot at in battle was something most people don't want to remember. All he had to do when he got out was walk across the street and sign some papers and he would have gotten a multi-engine commercial license, but he didn't do that.

After about 2 years of coming to the airport and watching and talking. He decided to fly again. We had a problem. I felt he needed to go through the whole program to learn even though he had 1,500 hours. He still needed 40 hours, cross country time, and instruments. I got him ready for his solo cross country, signed him off and sent him on his way to Moberly.

In those days, students would have to find someone at the airport to sign their logbook to show they made it. Well, Jim P. landed at Moberly and taxied in to the ramp. Unfortunately, he had too much coffee before he left and had a personal emergency. He ran into the lobby, threw his logbook on the desk and asked the attendant to sign his logbook.

When Jim P. came out, the guy said, "Are you kidding me?"

Jim P. said, "No, I'm on a student cross country. Is there a problem?"

Jim P. forgot his last cross country flight in his logbook had been from Hawaii to San Francisco.

The guy would hardly sign the book!

Dick was one of those people who sometimes had a brain drain. I solo-ed him and got him ready for his cross country. The first solo cross country for all the students was St. Clair to Hannibal, from Hannibal to Moberly, and then return. I liked to plan the cross countries so the student could leave St. Clair by about 10:00 a.m. and that would put them back by about 3:00 p.m. Three p.m. came, then four, then five, good thing it was the summer and didn't get dark until 8 p.m.

About 7:30 Dick calls in on the radio and said, "I made it!"

After he lands, he has gas tickets from all

over north Missouri, and even had two tickets from Mexico, Missouri. He's the only student who ever got all his cross country time in on one day! It seemed that he got lost, and then started following roads and water towers until he got home.

Since I was the main instructor, I had a lady named Valera who would fly the last three hours with my students and sign them off for the check ride. She started out in Vichy and gradually got all her ratings. She bought a Cessna 182 and then in the late sixties bought a Bonanza. She was a great help to me and probably had one of the best pass-fail ratios of anybody.

I always had the students ready for the check ride when she flew with them. Whenever she and her husband, Belmont, would come out to the airport to go on a trip. They would come in separate cars. But Belmont also had a commercial pilot's license. He was the one who didn't have his glasses on a check ride with Pete.

They owned a trailer sales and furniture store. They would fly people in from all over Missouri and Illinois to sell them a trailer and then sell them furniture. She said 75% of the people they flew in bought trailers.

Valera was in the Ninety-nines (women pilot's organization) and flew in many of the national races they held. Helping her get ready for the races was always fun, because we had to figure out how to make the bonanza go as fast as it could.

We used to tied the Bonanza down with large concrete blocks instead of real tie-downs, and it's a good thing we did.

Belmont came out to fly after she got in from a trip. I guess he decided he didn't need to preflight the airplane since she just got in. He untied the wing tie-downs but forgot the tail. He got in, started it up and taxied out for takeoff with the concrete block just bouncing on the asphalt. He turned around on the runway and gave it full throttle, accelerating down the runway; luckily, the block broke apart before he got airborne. If it had been a normal tie-down, then he would have ripped the tail of the aircraft right off.

In 1965, we decided to have an air show. We didn't want a big huge thing with a lot of acts and expensive airplanes. We figured we would just have airplane rides, sell cokes, and have parachute jumpers. By the end of the day, we had given 400 rides at $5.00 a piece and ran out of cokes. We thought we might have 100 rides and sell 3 cases of cokes.

We were just a little airport in the country, but I guess people wanted a little excitement.

The next year, we had a little experience and decided to throw a full blown air show. We started off with flag jumpers, aerobatics with a Citabria, a Gyrocopter, and then we had an act with a yellow cub.

Bill was a friend of my son Fred. They had gone to school together. Bill dressed up as a bum and started making a fool of himself as soon as the airshow started.

Then about an hour into the show, the

supposed pilot of the cub decided to give him a ride to get rid of him. He piles the bum, Bill, into the cub and since the cub didn't have a starter, the pilot had to hand propped the cub. The supposed pilot goes to the back and unties the tail wheel and the plane takes off without the pilot. Bill is taxiing down the runway. Then the announcer says to the crowd that someone is stealing the cub, and someone needs to do something about it.

Remember Harry? Well he was not in on the act. Harry jumps into the closest car and starts speeding down the side of the runway. He didn't drive on the runway cause he knew better. But the side of the runway was very rough and he was going about 60 mph and that old car was bouncing all over.

Bill was down at the end of the runway when Harry drove up, jumped out of the car, and stopped Bill from taking off. I don't know what Bill said to Harry, but finally, Bill takes off and does his act. That was the highlight of the show for all the performers. Sometimes Good Samaritans come entirely at the wrong time.

I did an occasional trip for a friend of mine. Basically, it was a charter but since we were friends, we called it flight instruction. The weather was not supposed to be very good. Cold and snow showers were forecast for all day. As was usual, Ken would sit and read the paper. The weather wasn't too bad on the way down, and I told Ken he better make his meeting short in Cape Girardeau which is about an hour flight from St. Clair.

Ken didn't make his meeting short, light snow showers had started, but it was still Visual Flight Rules (VFR.) VFR is when you aren't allowed to fly within the clouds. He finally got to the airport and we took off. We got about 10 miles from the airport and everything turned white. Visibility went to about 100 yards. The airplane wasn't rated for instruments, and I didn't have any charts with me so I decide to turn around and head back to the Cape. I called the tower and they said they were closed due to a snow storm. Well, we had no choice but to continue on toward St. Clair.

I normally flew VFR. So, it was nice to practice IFR for a little while. We continued on and began to wonder where the snow storm would end. We pass Farmington. I wasn't worried. We had plenty of gas to get us all the way to Iowa if need be. We got to Richwoods, a Visual Omni Range (VOR) station. It was still all white. A VOR station made up the highways in the sky.

I knew we were approaching St. Clair but hadn't seen anything since we flew into the storm. I knew we should be coming up to I-44 and St. Clair About the same time I knew we should be over the airport, the snow stopped, and there it was. We didn't have to fly to Iowa after all.

1968 comes around and we decide to outdo the last show. We contacted a radio station in St. Louis and made it the first annual radio station's air show. Thank goodness there wasn't a second annual. We went all out and the main attraction was having 2 of the D.J.s make their first parachute jump during the show. They supposedly didn't have any training, but they did.

We filled the airport with cars, airplanes, R.V.'s, you name it. I think we charged $5.00 a car load and nobody came in single. A conservative estimate was 25,000 people on the airport according to the paper. The airport was relatively small. The road to it was only a fourth of a mile long, and the fields around the airport were not much bigger. About the middle of the afternoon, the D.J.s were supposed to jump.

You have to realize that the St. Clair Airport is about 400 yards from I-44, an interstate 4 lane highway. We had so many people stopping on the highway, the highway patrol had to shoo them away. But the traffic was still slowed down to a crawl.

The D.J.s jump out of the airplane and their parachutes open, someone had misjudged the wind and the D.J.s were drifting toward the highway. The girl D.J. was about to land on the highway, right in front of a truck, luckily the truck wasn't going very fast. She lifted her feet trying to drift to the median and landed right on her butt in the middle of the road. But she was O.K. and the truck didn't hit her.

When the show was over, everyone left and headed to the local restaurants to eat. The restaurants all ran out of food! All we got to eat afterward were old hot dogs and leftover mustard. This was the last big show we ever had.

During this time, I was flying fire patrol to supplement our income. Missouri Fire Patrol didn't pay enough. So, I normally flew for the Federal Conservation Commission. The big problem with fire patrol was they didn't want to fly when there was no wind. They wanted to fly when most people would decide to stay on the ground.

The airports we went into were short, always crosswind, and some had a steep grade. Viburnum was one of those steep grade airports. You landed uphill and took off downhill, no matter the wind. Other airports were down in valleys or in between the trees with just enough wingtip clearance to miss the trees if you were in the middle of the runway.

One windy day, the feds called. I said it was too windy to go into the airport they wanted me to land and pick up someone. They said they didn't think I would go because they already called the state conservation who refused to go. They also called their own helicopter and he wouldn't go. But they were hoping I would.

Hilliard was one of the local pilots who owned a Cessna 185. There was no place Hilliard wouldn't go, because he had such a tough airplane. Hilliard left a little town in West Virginia one day after filling up and doing his business. His father-in-law was sitting in the right seat but didn't know anything about flying. Hilliard decided to take a nap after they get to altitude and level off.

Hilliard asked George to fly the airplane while he took a nap. As I said, George didn't know anything about flying, but he took over the controls anyway. Hilliard woke up about 3 hours later and asked George where they were. George said he didn't know. Hilliard started to look around and there they were, right over the airport they took off from 3 hours earlier. He never fell asleep on George again.

Hilliard also had a nasty habit of chewing tobacco. He usually had a can to spit in, in the airplane. He called me one day and said he had lost his keys and could I look in the airplane? One of the airport bums was in the office. He decided to look for them since I was busy. He looked all around the cockpit and finally decided the keys weren't there. There was one last place to look/ there was a bag hanging from the throttle.

He started to look in the bag and instead hits it with his hand kind of hard. The bag was t full of Hilliard's tobacco juice. It flew out all over this kid. What a mess. It did solve a problem though. The kid was more or less a pest, and he never came out again!

In 1971, I noticed that I couldn't see quite as well as I had, even glasses didn't help. When I would come in for a landing in the evening it looked like shooting stars were attacking me as I made my approach.

I went to the optometrist. But, he sent me to a specialist. It turned out I had a loose retina in my right eye. I thought my flying career was over. We scheduled a day for the operation. I was supposed to be in the hospital for three days. The strangest thing happened to me. They gave me a local anesthetic, and I was awake for the whole thing. The doctor took something like a spoon and popped my eye out of its socket. Then he very carefully put 11 stitches in the back of my eye. I could feel every one of those stitches, but it didn't hurt. I then proceeded to stay in the hospital room and listen to the news, ball games, whatever. The doctor told me I could look through my good eye, but only to keep my balance, nothing else. Well, when a good play would come on in baseball. They would show the instant replay. I would watch that!

Jim was going to school at Springfield, MO, at the time and came home for a week while I was in the hospital. On his way home, just as he made the turn to the airport, his engine seized up and ruined his car. Then he loaned out our tractor to the neighbors. They ruined one of the big tires and refused to repair it.

One of the pilots was trying to help out, but had an accident where he ran the Cessna 172 into a ditch and bent the propeller. All this happened after I went into the hospital and before I got home. Jim came up to get me at the hospital in St. Louis and had a car accident. It dented the car up a little but he wasn't hurt. He was an hour and a half late, and I was getting mad. Then he showed up and started telling me about his past week. I told him things can be repaired, as long as we were healthy. I believed that even though that week cost me a lot of money.

After I recovered and I thought everything was getting back to normal, Nan, my wife received the news that she had inoperable lung cancer, and she passed away the next month.

I was just about at the end of my rope

emotionally, it seemed like everything that could go wrong had. Then, as I was turning into the airport drive someone rear ends me. I go flying off the road. Luckily, I wasn't hurt but the car was totaled.

9 T-34 AND AEROBATICS

Well, after all that you can imagine my mental state. I became quite depressed. Jim was also going off to college to finish his last year at Southwest Missouri State (now Missouri State University.)

He came home every weekend, and I was appreciative of that. I always looked forward to his visits, and his work on the weekends. I tried to concentrate on the students, but I just couldn't get enthused about anything.

I even increased the size of my garden and decided to expand my plants hoping that would help. I planted about eight zucchini seeds and hoped that one would come up. I could tell my life was changing around the time all the zucchinis came up. That was the first time I ever planted squash and the plants came up. If you have ever had a zucchini plant, then you know the trouble I was about to get into.

The first plant put a gourd on, and I watched it grow. I ate it and enjoyed it, but then the other plants started putting on gourds upon gourds. I finally had my fill of zucchini; fried, in salad, boiled, and baked. You name it, I tried it. The zucchini kept coming. Finally, I started giving zucchinis to the students as they were leaving. Anybody who came to the airport got a zucchini.

Then one day, I went to town. I saw this pile of zucchinis on the side of the road where the customers had thrown out their zucchinis as they left the airport. That ditch still puts out zucchini plants every year!

A friend of mine Pete wanted his instrument rating so I instructed him. He just couldn't concentrate, that wasn't like him and I asked what was wrong. He said his teeth were bothering him and needed to have them fixed. I called off the training until he got all his teeth out. And son of a gun if he settled down and learned and got his instrument rating rather quickly. He just couldn't concentrate when he was in pain.

Jim went on and got his Certified Flight Instructor and Instrument Instructor (CFII), and Pete wanted him to get his multi-engine rating. Pete loaned us his Piper Twin Comanche. Pete thought that Jim's training wouldn't be complete without it. I felt he was right, but we didn't have a multi-engine airplane in the flight school. Jim would do just fine listening to me until he felt that he was good in the plane.

He made an approach to the airport, but wouldn't listen to what I was trying to get across. You know, that father and son thing. So he said let him make an approach his way.

I said, "O.K," just to see what happened.

He made a picture perfect approach and had the field made just right. Well, just as he was flaring for the approach I didn't ask, I told him, "Go around".

Well, we were under minimum controllable air speed (VMC) by about 10 mph, and I cut his left engine. He slowly brought the power up to a level which would keep us straight with maximum rudder. Then as we slowly accelerated he would increase the power a little more and we started climbing by the end of the runway. Well, that shook both of us up. I knew that if he made the mistake of slamming the power in we both would have been killed, and Pete would have been mad at us for wrecking his airplane. However, now I knew Jim would always handle an airplane in a safe manner.

My reputation was pretty good around Missouri, and one of the FBO's sent his son to fly with me for the commercial and instrument rating. The son wanted the best he could get, and he thought I was it. He went on to become an airline pilot.

Another airport, Rolla Downtown had instructors who were good, but one of the instructors wanted to get his instrument instructor's rating. Danny, the airport owner, told Robert he needed to fly at Rolla Downtown, but Robert wanted to fly with me. He flew 46 Nautical Miles to St. Clair, every time he wanted instruction for the rating. That really ticked off Danny, his boss. He was doing fine until he was almost ready for his check ride with the FAA. Robert was doing his pre-takeoff checklist and somehow missed setting the magnetic compass. We took off, and he couldn't fly the radial on the VOR. Then, he spotted his compass and knew right away what he had done. That was about the loudest "aw crap," I'd ever heard. After we landed, I told him, I thought he'd been ready for the checkride until he'd pulled that mess with heading indicator. Of course I didn't say it that nicely. Robert said, "Well Fred you could have told me about the heading indicator." I said, "If I did, you wouldn't remember to set it."

Several years later, he and his family came to the big open part of the hangar to eat lunch. They were talking about the airport and their trip. I didn't want to interrupt them. Then, the man comes inside and introduces himself and says, "Mr. Atkinson, I bet you don't remember me." I said, "sure I remember you. I bet you remember to set the heading indicator now too, don't you."

Later on, Robert called and wanted his Army Reserve flight instructors to come and meet me. I'm sure some of them were wondering what in the hell were they doing here at this little airport. I spoke to them about the responsibilities of flight instructing, not just the army way. It was quite a thrill for me to talk to those helicopter instructors,

I'm sure some of them flew in Vietnam, and

some may even have stayed in the army.

Students came and went. And life at the airport became a little boring for Jim and me. We heard about a program in the Civil Air Patrol (CAP) where they would sell their T-34's to the civilian market or maybe trade for a Cessna 172. We looked into it and the head of the CAP in Kansas City had heard of me instructing in T-34s 20 years earlier.

The Commanding Officer (CO) had 3 T-34s to trade or sell, and because we had kind of started the ball rolling the CO was saving the best one for us. Of course, the best one had 800 hours on the engine and the right brake didn't work. It was out of annual, had bad paint, had no radio, and needed about $5,000 in repairs. We traded a 1966 Cessna 172, in relatively good shape plus $551.00 for the T-34. As it turned out, this T-34 was one of the ones that I flew as an instructor during the war.

I flew the Cessna 172 over to Kansas City early one morning and was supposed to be home by 3:00 p.m. I got there and found out neither brake worked. So we had to fix at least one. That took about an hour. Then, I ran the engine and one magneto was out. That took a couple of hours. I had to take off from a controlled field and there was no radio so I called the tower on the telephone and got clearance for takeoff and departure.

Luckily, I remembered my light signals. I get airborne and headed for home. I didn't know if I should raise the landing gear or not. It might not come down! I raised it because I was late and didn't want to get home after dark.

I got back to St. Clair just before dark and buzzed the runway to wake everybody up. The landing gear came down, and I made a safe landing. Jim was really worried because it was 8:00 p.m. and I was supposed to be home four hours before. But it sure did feel great to be in a T-34 again. I took the T-34 over to Festus for the maintenance that needed to be done. That took about three months. There was a lot to fix.

During this time, I decided to make a big pot of chili. Jim liked to have a bowl when he came home from school. I always started my chili in a big stock pot, probably about a 2.5 gallon or so. I started by filling the pot half way up to the top with water and then adding chili powder, a good half a bottle. I liked my chili hot.

Then as its boiling, I'd brown my meat and start adding other stuff. I had just started boiling the water and chili powder when someone came for gas, then someone else came for something else, and somebody else came for a flight lesson. I finally get back to the kitchen about 3 hours later.

Red happened to be with me when I remembered the chili.

I rushed to the house with Red right behind me. I opened the door to a thick cloud of burned chili powder pouring out of the open door. The only thing I could think to do is hold my breath and run in and turn the stove off. I closed my eyes and took a deep breath and went into the trailer.

I felt my way to the stove, turned it off, and grabbed the pot with a rag and went out the back door. Red was waiting for me to come back. He was afraid I had collapsed or something. He started calling to me. I can't answer him from the back of the house because I'm trying to catch my breath. He stuck his head in, afraid of what he'd see. And I walk up behind him and scared him half to death. That's the last time I cooked without watching my food.

Eventually, we got the T-34 back and I start flying it about a half an hour at a time to work up to aerobatics. But oh did that feel good! It had been a long time since I had been doing aerobatics. After about three hours of practice I started on Jim. Now, we had something fun to fly, and Jim had come a long way from the early days of his learning to fly. Now he was a certified pilot.

Learning aerobatics made him a complete pilot now, and he'd never been afraid of unusual attitudes, or even being upside down.

We started getting more acclimated to the T-34 and started teaching other people to do aerobatics. This was the early 70's and not many people thought aerobatics were fun and easy to do. The next thing to do was have an airshow and show off what we had learned.

We called in some markers, got the pilots to volunteer some time and started planning our air show for 1976.

I think we had parachute jumpers, a hot air balloon, aerobatics, jets from the National Guard, helicopters from the National Guard, antique airplanes, and a new guy that had started hanging out at St. Clair. His name was Williard.

He had a Gyrocopter; it's sort of a helicopter and sort of a pusher airplane. He had built his gyrocopter and had traveled all over the country learning from the best gyro pilots.

Williard was good, no, great, with that gyrocopter. He had a routine that the crowd loved. We probably had about a thousand people and everyone had a good time.

One of student pilots from Weiss Airport in St. Louis called and wanted a check ride. This was very unusual since they didn't send me check rides very often. It turned out this lady's husband had come to me years ago, and she wanted to show him up. He had not passed his first time.

She came to the check ride and sure enough, she passed the first time. She wanted to learn more about flying so after this she started coming out about twice a week to start working on her instrument and commercial. She worked hard, asked hard questions, and made me think.

She was something I needed during this time. I was feeling lonely, and she made life interesting again. Not that there was a relationship there. It was more student and pupil. I was fairly happy again.

I made her an aerobatic pilot, got her commercial, CFI, and instrument ratings. She helped out instructing when Jim was not around, and she performed in our air shows. Word got around that I was a pretty good aerobatic instructor, and someone brought us a Cessna 150 aerobat to try out. I flew with the salesman and told him it was alright, but underpowered. I asked if I could take Doris up and see how she liked it. As we were climbing up to 2500 feet, I was explaining to Doris about the forward pressures required to maintain inverted level flight.

We started with an aileron roll, as we went inverted instead of easing the elevator forward as needed; she pushed the elevator forward, hard.

The nose went up. We had about a negative 3 G's on us and my head hit the skylight above my head and popped a perfect hole the size of my head in the skylight. I had blood pouring out of the cut around my head and Doris was screaming that she killed me.

We returned to the airport, and I just had a cut all the way around my head down to about my forehead. I knew what it was like to be scalped.

Doris helped me out a lot and my mental state began to improve after that terrible year. She always challenged me and made me think with all the weird questions she would ask. Then I noticed that I had a knot in my abdomen about the size of a pea, but it was growing.

I told Jim he had to come home, that I had to go to the hospital. He showed up the next day. I went into the hospital for an abdominal aneurysm. It could burst any minute, and I'd bleed out in about 2 minutes. By the time I entered the hospital, it was the size of a tennis ball.

The doctor thought that all those years of aerobatics and high G's may have contributed toward that. Maybe.

After this happened, I was not able to fly for about two months, but that didn't keep me from working in the garden, mowing on the tractor, all my usual chores around the airport. I must admit, I was waiting for my okay from the doctor by the time two months was up.

I wanted to get back to flying. After what seemed like two years I got my physical back. I couldn't wait to get back. I went up with Jim first, just to make sure I could get in and out of the Cessna 150 without any displeasure, and I was cleared to go.

GG, a local pilot, had a Cherokee 140 he kept at the airport. We really didn't know much about him. He was always friendly when he came in, but he never stayed around and talked to us. And we never got to fly with him. Then one day, he showed up with a broken jaw. Turned out he had gotten into a fight in a bar a few days before and the doctor told him to take a few days off from work.

He decided to learn aerobatics since he had to take the time off. He had his jaw wired shut and I was worried about him getting sick and drowning in his puke. He already solved the problem; he brought wire clippers with him.

That was the only time I included wire clippers in the checklist. He learned very fast, and we became great friends. It turned out he was a Green Beret, and the person he fought was badmouthing his unit. They said the other guy came out worse.

Later on, he bought a Taylorcraft and had it

rebuilt. It was beautiful with real nice nose art. He was real good with that airplane. One day after an annual he took off and noticed his wheel and tire came loose from the axle, and it had fallen to the ground during takeoff. How in the world was he going to land? He had been good in crosswinds with the Taylorcraft so he decides to just land on his right wheel.

He landed on his right main and tail wheel until his aileron became ineffective. Then he just stopped and turned a little. It was a beautiful landing.

Ernest was one of the local pilots who moved to St. Clair after he had his license. He heard stories about me. Some of his friends decided to fly down to Conway, Arkansas, for a meeting. The weather looked a little iffy so Ernest asked me to go along. I said sure, he could probably use a little dual along the way. Ernest showed up with all his maps, and his lubber line all drawn out to show the way.

He wanted to show me how experienced he was and how well he could navigate. Well, it started out very rough so Ernest asked me to fly and maybe make it a little smoother for the passengers. Okay, I could do that.

The flight did calm down a bit after that. He only had about 100 hours and didn't know how to fly smoothly for his passengers.

We had been flying about 45 minutes and he noticed that I didn't even have a map or tried to use the VOR. He was following along with his finger and roughly knew where he was. He didn't realize that I used to fly all over this country for fire patrol and knew every little town and road all the way to Arkansas.

He looked out and looked at the map and proudly announced that we were over Salem, Missouri. I whispered in his ear and said that was Houston, Missouri, about 15 miles away. He couldn't believe it so I proved it to him by flying low enough to see the water tower. He threw the map away.

Not long after this, Jim decided he wanted to move to South Texas. He set up his own flight school and banner towing service on the beach. Now, it was just Doris and me to take care of the students. I must admit, that was one of the better times in my life. Then about 3 years later, Doris got lung cancer.

I called Jim. He started making arrangements to move home.

10 THE CELEBRATION

Jim was not able to move until about June. But, he and Fred, Jr. gave me quite a surprise on April 1st, 1981. Jim and Fred drove up from Texas on March 30 to celebrate my fifty years of flying. I soloed on April 1st, 1931.

They were around and helping me out for a couple of days before. I was a little busy on April 1st, so I didn't notice the little things that were happening.

About 11:00 a.m, Betty and Wendell Tarman from the old days in Longview and Bonham just happened to come by and say hello. He and his wife lived close to Branson. They said they were on the way to St. Louis and thought they would stop in to say, "hi".

After a couple of students came by to fly, then Leroy Hoffman came by to talk about the old days of early St. Clair Airport. He was the pilot with the Cessna 170.

Then later that day, "Pete" Campbell came by and said he was working for the AOPA now and on his way to a safety meeting. He lived in Tennessee now and was the FAA man who told me I was to be an examiner. I didn't think anything about these coincidences. I was just glad to see them. Then about 6:00 p.m. my kids told me they were going to take me out to dinner and celebrate my 50 years.

I said "why do you want to do that?"

"Not many people achieve fifty years and they want to celebrate it. Dress up a little we're going to take you to a new place."

I guess I'm kind of naive and didn't think anything of it.

We drove up to the American Legion Hall.

"This must be a really good place, cause the parking lot is full."

We get out of the car and the boys go in before me. They prep the crowd. I go in the door and the first person I see is my daughter-in-law, Sandy, Jim's wife, and I began to think something was up.

I look beyond Sandy where people were starting to stand and applaud. What in the world? There's Betty and Wendell. There's "Pete" Campbell. There's Leroy. I can't count or remember all the names. Larry S., FAA. Carl W., FAA. Jim D., from Vichy. Most of the people in the room I had taught to fly or showed them a thing or two.

We have a dinner and I'm sitting at the daze right next to the podium. This must be in my honor.

A lot of people bring me gifts, cards, and memorabilia. Leroy starts the program off, and "Pete" ended it two hours later, many good memories and some wild stories were told but I had a great anniversary thanks to my kids and all the wonderful friends that came to celebrate with me.

Never in a million years would I have thought I would have had a party like this and no one told me anything about it, it was a total surprise.

The next day, it was back to normal. The kids left, and I went back to instructing. Jim and Sandy moved back in with me in June. They had been planning a trip to Mexico for a year and left for two weeks in August. When they got back, I had pneumonia and was sick for two weeks. Good thing they got back when they did. I probably wouldn't have survived another couple of days without them.

About that time, I find out that Valera had nominated me for National Flight Instructor of the Year. The competition for this honor is nationwide. First, I would have to win the St. Louis Region of the FAA. In September, I got good news. Sandy was pregnant with Jim and Sandy's first child. I was going to be a grandfather for the third time.

Oh, yes, I did win the competition also. Then I had to win the National competition. It turned out, there is a group of businessmen, aviation corporate executives, FAA, and AOPA who vote on the recipient they thought was best.

When it came time for the vote, I was the only candidate to ever receive votes from everyone. This was the first time that had ever happened. I was the 1981 National Flight Instructor of the Year.

I took the trip to Washington, went through the Smithsonian, toured Washington, met a lot of people and had a good time. They showered me with gifts and money. Wow, what a year!

I got back to the airport and continued my daily routine. However, I'm 70 years old, and starting to feel it. My flying ability is not affected, but my teaching ability is suffering. I used to begin teaching students and once I had them soloed and ready for the dual cross country, then I would turn them over to Jim and he would finish them and get them ready for the check ride.

Now, I feel that I'm not doing as well with the students and the check rides. So, I tell Carl W. at the FAA that I wanted to resign from the examiner system, but I would like to recommend Jim for my position. He heartily agreed. Jim went to Oklahoma City for a week's worth of training. He passed, and we traded places in the curriculum.

Jim wanted to make some changes around the airport, to save us money, and increase business. He suggested we put in auto fuel for our airplanes. We built a new ten unit hangar. Things were starting to look up. We approached the city about building an office and terminal building. They said there is no money. That seemed to be the only thing they could say.

In the meantime, we were teaching and greeting customers in the shabby office in the back of the hangar we have had for 20 years. About once a year, sometimes twice, the creek next to the hangar seemed to flood our office and made it a little bit dirtier. We couldn't keep nice furniture. It got ruined every year. We had flood lines on the wall from each of the floods and then the city decided it was time to expand the airport.

They didn't use any of the money for a terminal building. They expanded the runway and made a taxiway. We needed that, but they could have put money into the expansion for other things or at least upgrade the next year. No, they didn't have the money.

So, the next year, we got a call from the city secretary about 4:00 p.m. to come to the city council meeting at 7:00 p.m. We asked why and they said that we'd find out. Well, Jim and I wondered what is going on. They had never asked us to council meeting. We get there and the first thing on the agenda is the airport.

The mayor said they have gotten many reports about the airport being in shabby condition and dirty all the time.

Jim got up and said, "We have been fighting flooding for five years and get as much as six inches of flood water in the office every year. Even Mike, the council member has seen this. We have asked for a better office many times to get out of the flood plain but are always answered with, 'No money.'"

We got home and talked things over. Finally, we decided the city wanted us out. We had done the best we could for the city, paid our rent on time every month, brought many new airplanes to the airport, brought a lot of outside money into the city, brought national recognition to the airport and helped to expand the airport. All while they did nothing.

So, we decided to resign effective three months later. We liquidated things a little at a time, trying to get any amount of money out of 24 years of running an airport. We sold the T-34, one of the Cessna 150's and that continued on for a little while longer. We figured we'd rent a house in St. Clair and continue to fly in the other airplanes. Then, Sandy got a call from Port Aransas, Texas, to come and teach.

We got rid of everything. We made our trek to Port Aransas, a tiny fishing town on an island off the coast of Corpus Christi, Texas, It was 1,000 miles away. We found a house, talked to the FBO, got flying jobs, and settled in to a new life. It was time to start a new adventure, and what a better way to do it then to be on a coast.

ABOUT THE AUTHORS

Fred, Jr. Atkinson received his Commercial Pilots License from his father and enjoyed flying for four years before focusing on his career. He worked at Mcdonnell and Vought until his death in 2001.

James Atkinson also received his Commercial Pilots License from his father. He became a flight instructor, instrument flight instructor, multi-engine flight instructor, and aerobatic instructor.

Fred Atkinson Senior passed away in 1988.

James quit flying for ten years after his father's death until it was time to return.

The stories in this biography were passed down from father to sons during Fred Sr. lifetime.

Made in the USA
Middletown, DE
31 March 2022